Determination of Air Quality

Determination of Air Quality

Proceedings of the ACS Symposium on Determination of Air Quality held in Los Angeles, California, April 1-2, 1971

Edited by
Gleb Mamantov
Department of Chemistry
University of Tennessee
Knoxville, Tennessee

and

W. D. Shults
Analytical Chemistry Division
Oak Ridge National Laboratory
Oak Ridge, Tennessee

SPRINGER SCIENCE+BUSINESS MEDIA, LLC

Library of Congress Catalog Card Number 72-182861
ISBN 978-1-4615-8665-4 ISBN 978-1-4615-8663-0 (eBook)
DOI 10.1007/978-1-4615-8663-0

Originally published by Plenum Press, New York in 1972
Softcover reprint of the hardcover 1st edition 1972

PREFACE

In arranging for this Symposium on the Determination of Air Quality, we attempted to present much more than analytical chemical information. We realized fully that much of the audience would be interested in that type of information, but we also believed strongly that these analytical chemists want and need to know the fate and significance of their products, *i.e.*, their data. Accordingly, the participants were drawn from numerous "disciplines" -- administrators, medical researchers, engineers, systems analysts, and instrumental and chemical analysts. There was a corresponding diversity of subject matter within the formal presentations.

The Symposium was conducted in three half-day sessions. The first of these addressed the general subject of what is being done now regarding the determination of air quality. This general subject touched upon present data storage and handling activities, surveillance networks, correlative work with health effects, and efforts to combine (or index) several measured parameters into a single understandable value. The second session dealt with recent developments in the analytical methodology of air quality. Research and review papers were presented. The final session addressed more *avant garde* topics, such as the determination of odors, the use of electron spectroscopy for air quality studies, and the important intersociety effort aimed at standardizing analytical procedures in the air quality area.

The response to this symposium was gratifying. It strengthened our belief that topical symposia can go far to improve communication between people who are attacking a common problem, but from different directions.

In the interest of rapid publication, each author supplied a typed manuscript ready for photo-offset duplication. Editorial manipulation was minimized. We believe that rapid and enhanced multidisciplinary communication of information through symposia of this type is valuable. It means that the value of the symposium is greater than the value of the sum of the individual pieces.

Gleb Mamantov
University of Tennessee
Wilbur D. Shults
Oak Ridge National Laboratory

CONTRIBUTORS

Lyndon R. Babcock, Jr., Department of Energy Engineering, University of Illinois at Chicago Circle, Chicago, Illinois

Lewis F. Ballard, Research Triangle Institute, Research Triangle Park, North Carolina

Delbert S. Barth, Environmental Protection Agency, Air Pollution Control Office, Bureau of Air Pollution Sciences, Durham, North Carolina

F. Benson, Community Research Branch, Division of Effects Research, Bureau of Air Pollution Sciences, Office of Research and Monitoring, Environmental Protection Agency, Durham, North Carolina

D. C. Calafiore, Community Research Branch, Division of Effects Research, Bureau of Air Pollution Sciences, Office of Research and Monitoring, Environmental Protection Agency, Durham, North Carolina

T. A. Carlson, Oak Ridge National Laboratory, Oak Ridge, Tennessee

Clifford E. Decker, Research Triangle Institute, Research Triangle Park, North Carolina

E. R. de Vera, Air and Industrial Hygiene Laboratory, California State Department of Public Health, Berkeley, California

Andrew Dravnieks, Odor Science Center, IIT Research Institute, Chicago, Illinois

J. L. Durham, Environmental Protection Agency, Air Pollution Control Office, Durham, North Carolina

J. F. Finklea, Community Research Branch, Division of Effects Research, Bureau of Air Pollution Sciences, Office of Research and Monitoring, Environmental Protection Agency, Durham, North Carolina

B. R. Fish, Oak Ridge National Laboratory, Oak Ridge, Tennessee

G. G. Guilbault, Department of Chemistry, Louisiana State University, New Orleans, Louisiana

M. Guyer, Research Triangle Institute, Research Triangle Park, North Carolina

D. I. Hammer, Community Research Branch, Division of Effects Research, Bureau of Air Pollution Sciences, Office of Research and Monitoring, Environmental Protection Agency, Durham, North Carolina

E. R. Hendrickson, Environmental Engineering, Incorporated, Gainesville, Florida

T. A. Hinners, Community Research Branch, Division of Effects Research, Bureau of Air Pollution Sciences, Office of Research and Monitoring, Environmental Protection Agency, Durham, North Carolina

J. A. Hodgeson, Environmental Protection Agency, Technical Center, Research Triangle Park, North Carolina

S. Hsiung, Department of Chemistry, Louisiana State University, New Orleans, Louisiana

L. D. Hulett, Oak Ridge National Laboratory, Oak Ridge, Tennessee

W. Hussein, Department of Chemistry, Louisiana State University, New Orleans, Louisiana

Julian F. Keil, Medical University of South Carolina, Charleston, South Carolina

E. L. Keitz, The MITRE Corporation, McLean, Virginia

S. S. Kuan, Department of Chemistry, Louisiana State University, New Orleans, Louisiana

O. Menis, Analytical Chemistry Division, National Bureau of Standards, Washington, D. C.

Clinton Miller, Medical University of South Carolina, Charleston, South Carolina

T. R. Mongan, Sydney Area Transport Study, Sydney, Australia

G. B. Morgan, Environmental Protection Agency, Air Pollution Control Office, Bureau of Air Pollution Sciences, Division of Atmospheric Surveillance, Raleigh, North Carolina

P. K. Mueller, Air and Industrial Hygiene Laboratory, California State Department of Public Health, Berkeley, California

W. Nelson, Community Research Branch, Division of Effects Research, Bureau of Air Pollution Sciences, Office of Research and Monitoring, Environmental Protection Agency, Durham, North Carolina

V. A. Newill, Community Research Branch, Division of Effects Research, Bureau of Air Pollution Sciences, Office of Research and Monitoring, Environmental Protection Agency, Durham, North Carolina

C. Pinkerton, Community Research Branch, Division of Effects Research, Bureau of Air Pollution Sciences, Office of Research and Monitoring, Environmental Protection Agency, Durham, North Carolina

T. C. Rains, Analytical Chemistry Division, National Bureau of Standards, Washington, D. C.

Edgar A. Rinehart, Physics Department, University of Wyoming, Laramie, Wyoming

T. A. Rush, Analytical Chemistry Division, National Bureau of Standards, Washington, D. C.

M. H. Sada, Department of Chemistry, Louisiana State University, New Orleans, Louisiana

Samuel H. Sandifer, Medical University of South Carolina, Charleston, South Carolina

C. R. Sawicki, Research Triangle Institute, Research Triangle Park, North Carolina

E. Sawicki, Research Triangle Institute, Research Triangle Park, North Carolina

F. Scaringelli, Raleigh, North Carolina

C. M. Shy, Community Research Branch, Division of Effects Research, Bureau of Air Pollution Sciences, Office of Research and Monitoring, Environmental Protection Agency, Durham, North Carolina

Arthur C. Stern, University of North Carolina, Chapel Hill, North Carolina

Robert K. Stevens, Environmental Protection Agency, Technical Center, Research Triangle Park, North Carolina

E. C. Tabor, Environmental Protection Agency, Air Pollution Control Office, Bureau of Air Pollution Sciences, Division of Atmospheric Surveillance, Raleigh, North Carolina

R. J. Thompson, Environmental Protection Agency, Air Pollution Control Office, Bureau of Air Pollution Sciences, Division of Atmospheric Surveillance, Raleigh, North Carolina

Y. Tokiwa, Air and Industrial Hygiene Laboratory, California State Department of Public Health, Berkeley, California

SuzAnne Twiss, Air and Industrial Hygiene Laboratory, California State Department of Public Health, Berkeley, California

Philip W. West, Coates Chemical Laboratories, Louisiana State University, Baton Rouge, Louisiana

CONTENTS

Atmospheric Surveillance--Past, Present, and Future 1
G. B. Morgan, E. C. Tabor, and R. J. Thompson

Importance of Air Quality Measurements to Criteria, Standards, and Implementation Plans 19
D. S. Barth

Aerometric Data: Needs and Networks 25
T. R. Mongan and E. L. Keitz

A Program of Community Health and Environmental Surveillance Studies (CHESS) 41
C. M. Shy, J. F. Finklea, D. C. Calafiore, F. Benson, W. Nelson, and V. A. Newill

Human Pollutant Burdens 49
J. F. Finklea, D. I. Hammer, T. A. Hinners, and C. Pinkerton

Pesticide Exposure Index (PEI) 57
J. E. Keil, J. F. Finklea, S. H. Sandifer, and M. C. Miller

A Combined Index for Measurement of Total Air Pollution: Effects of Changing Air Quality Standards 65
L. R. Babcock, Jr.

Ratio of Sulfur Dioxide to Total Gaseous Sulfur Compounds and Ozone to Total Oxidants in the Los Angeles Atmosphere--An Instrument Evaluation Study 83
R. K. Stevens, J. A. Hodgeson, L. F. Ballard, and C. E. Decker

Atmospheric Ozone Determination by Amperometry and Colorimetry 109
Y. Tokiwa, S. Twiss, E. R. de Vera, and P. K. Mueller

The Determination of Trace Metals in Air 131
P. W. West

The Activities of the Intersociety Committee on Manual of Methods for Ambient Air Sampling and Analysis 143
A. C. Stern

The Status of Sensory Methods for Ambient Air Monitoring 155
E. R. Hendrickson

Interfacing of Sensory and Analytical Measurements 163
A. Dravnieks

Studies of Sulfur Compounds Adsorbed on Smoke Particles and Other Solids by Photoelectron Spectroscopy . . . 179
L. D. Hulett, T. A. Carlson, B. R. Fish, and J. L. Durham

Comparison of Methods for the Determination of Nitrate--Determination of Nitrate Through Reduction (Abstract) 189
C. R. Sawicki and F. Scaringelli

Photometric Determination of Polyphenols in Particulate Matter (Abstract) 190
E. Sawicki and M. Guyer

An Evaluation of Atomic Absorption and Flame Emission Spectrometry for Air Pollution Analysis (Abstract) 191
T. C. Rains, T. A. Rush, and O. Menis

Analysis of the Aerocarcinogen Conglomerate (Abstract) . . . 192
E. Sawicki

Microwave Spectrometry as an Air Pollutant Analysis Method (Abstract) 193
E. A. Rinehart

Biological Degradation of Toxic Pollutants (Abstract) 194
G. G. Guilbault, S. S. Kuan, M. H. Sada, W. Hussein, and S. Hsiung

Index 195

ATMOSPHERIC SURVEILLANCE--PAST, PRESENT AND FUTURE

G. B. Morgan, E. C. Tabor and R. J. Thompson

ENVIRONMENTAL PROTECTION AGENCY
Air Pollution Control Office
Bureau of Air Pollution Sciences
Division of Atmospheric Surveillance

Atmospheric surveillance is defined as the systematic collection and evaluation of aerometric data, which includes information on pollutant emissions, pollutant concentrations and meteorological parameters.

Information provided by an adequate atmospheric surveillance system in a polluted area initially demonstrates the need for control of air pollution; thus it provides the basis for control plans and finally indicates the effectiveness of control measures.

An ambient air surveillance system may provide on-the-spot measurements of pollutant concentrations or may collect samples for subsequent analysis. Some systems may combine the two techniques, depending on the pollutant to be measured and the information needed.

From the beginning of the fourteenth century in Great Britain, air pollution resulting from the burning of soft coal was recognized and many attempts made to solve the problem (1). During this period surveillance was a private matter and was no doubt quite subjective, because the only sensors used were human eyes and noses.

By the beginning of the nineteenth century, the smoke nuisance in London was of such magnitude that a Select Committee of the British Parliament was appointed in 1819 to study and report on smoke abatement. The accomplishments of this committee are not documented.

In the latter part of the nineteenth century, similar conditions had developed in the United States. Chicago and Cincinnati enacted smoke control regulations in 1881. Similar regulations were effected by 1912 in 23 of the 28 cities with populations over 200,000.

EARLY AIR QUALITY SURVEILLANCE

The Beginning

As a result of growing concern over the air pollution problem, many cities initiated elementary surveillance programs in the 1930's and 1940's. A few cities acted even earlier: Chicago conducted a detailed evaluation of that city's problem around 1920. The U. S. Public Health Service (2) reported on air pollution in American cities during the years 1931 to 1933.

In a summary of early air surveillance prepared in 1956, Wohlers and Bell (3) reported that at least 21 cities had either conducted or were conducting dustfall measurement programs. Earliest reports of these measurements were those of Pittsburgh (1912-13) and Salt Lake City (1919-20) followed by reports from Baltimore, Cincinnati, Cleveland, Detroit, and St. Louis in the late 1920's or early 1930's. Salt Lake City and Pittsburgh pioneered in sulfur dioxide measurements, later joined by 42 other cities. Documentation is lacking on the history of sulfation rate measurement in the United States. Before the early 1950's, thirteen cities made sporadic measurements for one or more of several gaseous pollutants--H_2S, NOx, NH_3, CO, fluorides, oxidants, hydrocarbons, and aldehydes (Table I).

Surveillance in Los Angeles

During the late 1940's and early 1950's, a tremendous effort was expended to precisely define the nature of the air pollution problem in the Los Angeles area. An analysis of the complexity of the problem, the large area involved, and the meteorology of the area resulted in development of a system for the continuous surveillance of several pollutants at several sites in Los Angeles County (4,5,6). Twelve stations strategically located throughout the county were established late in 1954, equipped with automatic instruments that sampled and analyzed the pollutants on a continuous basis. Pollutants measured eventually included CO, total oxidants, oxidant precursors, NO, NO_2, SO_2, hydrocarbons, and particulate matter. These events mark the beginning of the first effective systematic and routine atmospheric surveillance program.

TABLE I

EARLY MEASUREMENT OF GASEOUS POLLUTANTS

City	H_2S	NOx	NH_3	ald.	HC	CO	Ox
Baltimore		+	+	+		+	+
Charleston	+	+	+	+			
Chicago			+				
Cincinnati		+	+	+		+	
Detroit						+	+
Donora		+				+	+
Long Beach					+		
Los Angeles		+	+	+	+	+	+
Louisville				+			
Pittsburgh	+						
Salt Lake City			+				
San Diego		+		+	+	+	+
Santa Clara	+	+		+		+	+

The National Air Sampling Network (NASN)

In 1953 the Sanitary Engineering Center of the Public Health Service, under a contract with the Army Chemical Corps, embarked on a program to measure the protein content of the air. This effort required development of a suitable sampling system to collect sufficient suspended particulate matter to meet analytical requirements. A recently developed, highly efficient glass fiber filter was combined with a modified Silverman air mover; this was the forerunner of the present Hi-Vol glass fiber filter sampling system. Development of analytical methodology progressed in parallel with that of the sampling system.

From the beginning the network was operated as a cooperative endeavor. Equipment and supplies were provided by the Public Health Service, and sampling was done by the local health department or other agency personnel. Beginning with two stations in September 1953, the network expanded in a random fashion to a maximum of 64 stations (urban, suburban, nonurban) by December 1956, as shown in Figure 1 (7).

Figure 1. National Air Sampling Network - 1956

The unorganized increase in size of the network was the result of a desire to comply with many requests from local agencies to participate during this period. It became obvious that a more systematic operation would better meet the need for a complete definition of the nation's air quality. As a result, in 1956, a master plan for orderly development and operation of a NASN was developed, to become effective in 1957 (8). The plan imposed a limit of one sampling site per city and replaced residential or suburban sites with nonurban sites as remote as possible from urban influences. The new plan provided for surveillance of all the major population centers and for one nonurban sampling station in each state. Thus came into existence a truly national air surveillance program. The number of urban stations has varied from year to year, averaging about 250 in recent years.

From the beginning, every effort was expended to obtain the maximum amount of information from each sample. Accordingly, the samples were analyzed for suspended particulate matter, benzene-soluble organics, water-soluble nitrates and sulfates, lead, beryllium, nickel, chromium, vanadium, and several other metals. In addition, many samples were analyzed for fluorides, arsenic, ammonium, and benzo(a)pyrene. In the early years, before the establishment of the Radiation Alert Network, measurements of gross beta radioactivity were made on all samples. These measurements provided national coverage during the era of high nuclear activity (9). In 1959, the NASN embarked on a sampling program for SO_2 and NO_2, operating a bubbler-type sampler that was developed specifically for network use and provided 24-hour average concentrations (10). This gas sampling network was gradually expanded to the present total of 208 stations, which are widely distributed. The NASN gas sampler has proved highly useful for collection of integrated samples of numerous gaseous pollutants.

Rapid expansion of the NASN program increased tremendously the quantity of air quality data produced. The burden of data processing and analysis grew so heavy that it early became obvious that more modern data handling techniques must be adopted. Consequently, with the implementation of the modern network in 1957, electronic data processing became an integral part of the program. This initial effort was the first directed toward efficient routine handling and analysis of data generated on a large scale by an air surveillance network.

In cooperation with NOAA (formerly U. S. Weather Bureau) a national precipitation collection network of 29 stations was established in 1960 and operated through 1964. Monthly samples of precipitation were analyzed for several pollutants. The data provided new information about washout and transport of pollutants.

Although analytical capabilities were developed as the network's sampling activities expanded, input from the gas sampling network imposed a greater load on the laboratory than could be handled by the usual manual methods. This problem was partially resolved by adoption of automated techniques for routine analyses whenever possible. Ultimately the laboratory instrumentation was improved to the extent that all routine analyses for gaseous pollutants, water-soluble ions, and trace metals were automated.

A major event in national air surveillance was establishment of the continuous air monitoring program (CAMP) in 1961 (11). CAMP stations were equipped with instrumentation capable of measuring continuously and automatically the ambient concentrations of CO, SO_2, NO, NO_2, total oxidants, and total hydrocarbons. Initially the stations were established in six cities: Chicago, Cincinnati, New Orleans, Philadelphia, San Francisco, and Washington. This program was initiated to provide a detailed definition of the concentrations and characteristics of gaseous pollution in urban areas of widely differing characteristics. Over the years, this program has generated tremendous amounts of data relative to gaseous pollution, provided training for State and local air pollution control personnel, stimulated the establishment of similar programs by State and local agencies, and helped to inform the increasingly concerned public about air pollution.

The product of any surveillance system should be presentation of the data in such a manner that the results may be easily understood and interpreted and that the original objectives are satisfied. The NASN has attempted to do this by publishing and distributing numerous reports of varying degrees of sophistication. The need to process, analyze, and disseminate vast amounts of data led to the development of SAROAD (12), a system for the automated storage and retrieval of aerometric data.

State and Local Networks

Cooperation of State and local agencies in the early NASN programs stimulated considerable interest within these agencies. As a result, several States established air surveillance networks to define local air quality more fully. Among the first States to operate such programs were Maryland, Massachusetts, New Jersey, New York, Texas, and Washington. These State networks adopted sampling and analytical procedures that were comparable in every way with those of the NASN. NASN staff provided on-the-job training in NASN laboratories for state personnel and also assisted by visiting many State agencies to review their sampling and analysis methodologies. The NASN further supported certain State programs by providing sampling equipment and laboratory services until their own State governments could provide the required resources.

With the enactment of Federal legislation in 1963, financial assistance became available from the Federal Government. This increase in support was reflected within a short time in a steady and substantial growth in State and local air surveillance activities. More agencies have established surveillance programs, and those with ongoing programs have expanded them both in area coverage and in sophistication of instrumentation and techniques.

CURRENT NATIONAL SURVEILLANCE

The Air Quality Act of 1967 underlies much of the current pollution control effort. This Act initiated the enforcement of air quality standards on a regional basis. Adequate air surveillance systems are required as part of implementation plans designed to assure compliance with air quality standards. The primary objectives of atmospheric surveillance are: 1) to gather information on the ambient levels and trends in levels of numerous pollutants, both gaseous and particulate, 2) to gather information on the levels and trends in levels of selected pollutants--solids, liquids, and gases--in rural or "background" areas, 3) to identify and measure new or newly recognized pollutants, 4) to increase the basic understanding of source-receptor relationships, atmospheric interactions, and pollutant interrelationships, and how these complex relationships may be expressed in mathematical models, 5) to provide a basis for evaluating compliance with or progress made toward meeting ambient air quality standards, 6) to activate emergency control procedures during air pollution episodes, and 7) to provide a data base for use in evaluating pollutant effects in urban and transportation planning, and in developing and evaluating control strategies.

Current knowledge of atmospheric pollutants and their sources ranges from a large body of information about several pollutants (total suspended particulates, SO_2, dustfall) to sparse or nonexistent information about others. Only about 25 Air Quality Control Regions in the country can provide adequate information on CO and total oxidants (O_3).

Present-day surveillance of the nation's air quality is an integrated effort involving local, regional, State, and Federal air pollution control agencies. Such a program is necessary to determine the extent of people's exposure to air pollution and to assess the impact of air pollution control measures. Today some 7000 pollutant monitoring devices are in routine operation--approximately 3500 static devices, 2400 mechanical (integrating devices), 460 AISI spot tape samplers, and approximately 235 automatic (continuous) sampler/analyzers (13).

The surveillance programs conducted by State and local agencies are directed toward enforcement activities and designed primarily to sample the atmosphere for pollutants for which the Environmental Protection Agency (EPA) has issued criteria documents: particulate matter, SO_2, NO_2, CO, photochemical oxidants, and hydrocarbons. An adequate surveillance system will provide timely and valid data on specific pollutants to determine whether their concentrations exceed the standards, to indicate what control actions are needed, to determine air quality in nonurban areas of a region, and to allow control officials to evaluate deterioration of airquality during air pollution episodes.

The main goal of the Federal Government in atmospheric surveillance is to collect and make available requisite aerometric and related data on a nationwide scale. It is the intent of the Federal program, now and in the years to come, to utilize its efforts and resources to develop a unified nationwide program of surveillance through operation of the National Aerometric Data Bank (NADB) and through coordination of Federal, State and local programs. Where State and local agencies perform the primary surveillance, the Federal effort is aimed at collecting and validating the data that enter the National Aerometric Data Bank from nonfederal sources.

Ambient air quality standards are not scheduled for certain pollutants; the primary responsibility for surveillance remains with the Federal Government. This group includes (1) those for which national emission standards will be forthcoming, (2) those that are suspected of implication in health and welfare effects, and (3) those of interest because of their interactions in the atmosphere.

To carry out this responsibility, the Federal Government must operate several networks for the measurement of numerous pollutants, both particulate and gaseous. During calendar year 1970, EPA operated several sampling networks throughout the nation. Figure 2 shows sampling site locations for the Hi-Vol total suspended particulate network. Figure 3 shows the total mercury network; Figure 4, the membrane filter particulate and precipitation networks.

The atmospheric pollutants that EPA is currently measuring are listed in Table II (14). For some of these pollutants, a great deal of data are available; for others, information is limited. For pollutants such as mercury, selenium, and pesticides, methodology must be improved or developed to permit accurate ambient measurements on a network basis.

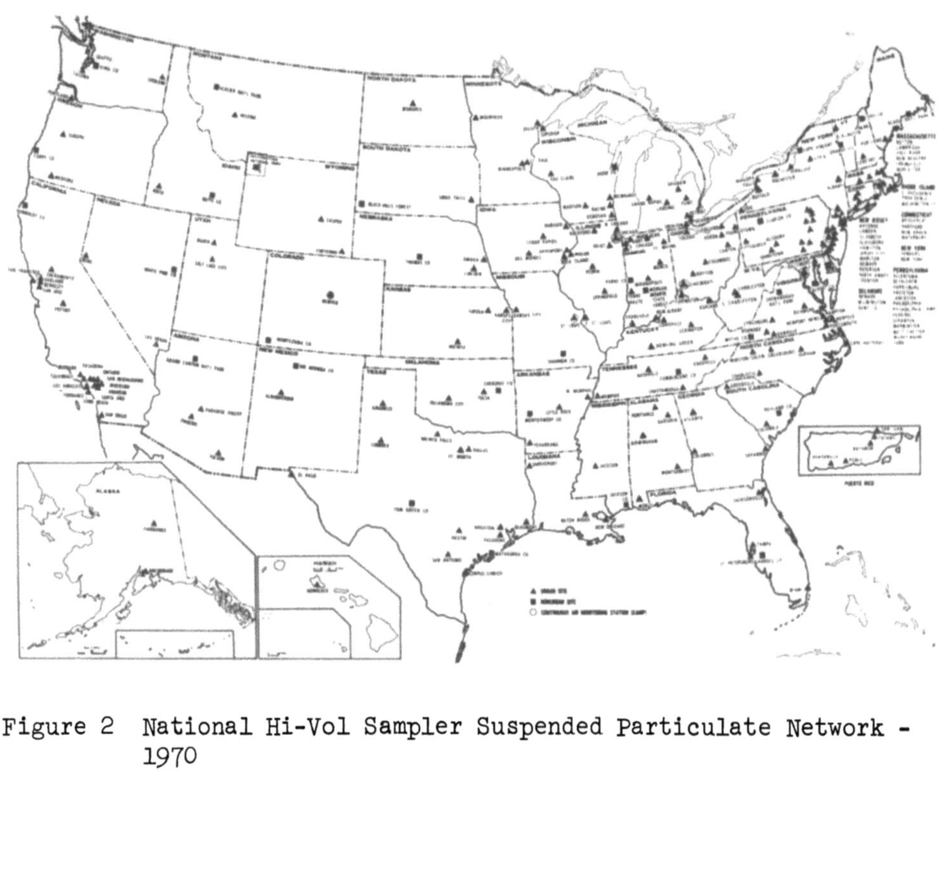

Figure 2 National Hi-Vol Sampler Suspended Particulate Network - 1970

Figure 3 National Mercury Collection Network - 1970

Figure 4 National Membrane Filter Particulate and Precipitation Networks

TABLE II

ATMOSPHERIC POLLUTANTS CURRENTLY MEASURED

TOTAL SUSPENDED PARTICULATE MATTER

Constituents of Total Suspended Particulate Matter

Inorganics - as total element or radical

Antimony	Cobalt	Tin
Arsenic	Copper	Titanium
[a]Barium	Iron	Vanadium
Beryllium	Lead	Zinc
Bismuth	Manganese	Ammonium
[a]Boron	Molybdenum	Fluoride
Cadmium	Nickel	Nitrate
Chromium	[a]Selenium	Sulfate

Organics

Benzene-soluble organic compounds
Benzo(α)pyrene
Benzanthrone

GASES

Sulfur dioxide	Carbon monoxide
Total oxidants (O_3)	Methane
Nitric oxide	Total hydrocarbons
Nitrogen dioxide	Nonmethane hydrocarbons

MISCELLANEOUS

[a]Asbestos	[a]Pesticides (solids plus vapors)
β radioactivity	[a]Mercury (solid, vapor, organic)
[a]Aeroallergens	[a]Suspended particulate size distribution

[a]Method development, evaluation, and pilot measurement program underway.

SURVEILLANCE IN THE FUTURE

The National Aerometric Data Information Service

The Clean Air Amendments of 1970 will increase the need for aerometric data by agencies on all levels. To meet these requirements, EPA is implementing a system to accelerate, expand, and coordinate the collection and dissemination of aerometric data throughout the nation. This is the National Aerometric Data Information Service (NADIS), a systems approach to collecting, storing, and retrieving data to provide an orderly and timely flow of aerometric data to and from the National Aerometric Data Bank (NADB).

To be useful, NADIS must produce data and information that are summarized, structured, and presented in a way that permits maximal utilization by control officials, researchers, industrialists, and other interested citizens. To develop such an information service, EPA is considering three basic goals: 1) Surveillance networks must be operated throughout the various regions of the country. In line with this goal, EPA has recently published a document; "Guidelines: Air Quality Surveillance Networks," (15) to assist State and local agencies. 2) Data obtained from surveillance efforts must be accurate. EPA's "Guidelines for the Acquisition of Validated Air Quality Data" (16) describes calibration techniques for field instrumentation and standardization of laboratory operations as an approach to improving data quality. 3) Air quality data must be current. EPA's SAROAD Users Manual (17) has been prepared to guide State and local agency personnel in the timely coding and processing of their surveillance data for entry into the data bank.

The type of information system appropriate for a particular area is determined by such factors as the potential for episodes of high pollution, potential severity of such episodes (depending on the concentrations and types of industry within the region), population of the area, size of the agency, and personnel available to operate and service the system. NADIS information systems are classified and characteristics of each class presented in Table III. During the past year, EPA and its contractors have contacted some 245 air pollution control agencies to obtain information on their present information systems and the numbers and types of sampling systems being operated. Table IV indicates the numbers of State and local agencies that currently comprise each of the five NADIS classes.

Figure 5 depicts an overview of the National Aerometric Data Information Service. An EPA computer center functions as the hub

TABLE III

CLASSIFICATION OF NADIS INFORMATION SYSTEMS

		Data Transmission (Highest Order)		
Class	Instrument Characteristics	To APCO From SAQIS[a]	To SAQIS[a] From Local	To Local From Sampling Site
I	Automated	Dataphone	Dataphone	On-Line Telemetry
II	Automated	Dataphone	Dataphone	Manual Retrieval: (Dataphone Used During Episode)
III	Automated and/or Mechanical	Dataphone	Dataphone	Manual Retrieval: (Warning System Used During Episode)
IV	Automated and/or Mechanical	Dataphone	Mail	Manual Retrieval
V	Mechanical	Dataphone	Mail	Manual Retrieval

[a]State Air Quality Information Service

TABLE IV

PRESENT STATE/LOCAL INFORMATION SYSTEMS BY NADIS CLASS

Total Air Pollution Agencies	245
Class I	12
Class II	9
Class III	0
Class IV	74
Class V	126
Unknown	2
Do Not Operate A System	22

of the operation. Routinely, the data flow is from local to State agency, then to the National Aerometric Data Bank. During potential or actual air pollution episodes, however, data can flow directly between the State and local surveillance systems and between these systems and the data bank. This overview suggests some of the versatility of the NADIS system, which can accomplish storage and retrieval of data on air pollutant concentrations, sources, and emissions; on meteorological conditions; or on any other environmental parameter. The system is flexible enough for use in any similar type of operation.

Presently NADIS is structured for on-line rapid data acceptance by dataphone or equivalent. It is further structured so that it can be queried in terms of geographical area, pollutant, or time period to respond to specific needs. Further, the system can accept and store data from more than ten thousand sites for approximately 30 parameters from each site. Input from local agencies can be by dataphone, magnetic tape, punched cards, or paper forms. The processes by which data can be deposited into and retrieved from the National Aerometric Data Bank are illustrated in Figure 6. Currently, the bank contains over four hundred million characters, and projections are that by 1976, when NADIS is scheduled to be fully operational, the bank will incorporate more than five billion characters.

Advances in Measurement Techniques

As indicated earlier, the methods for sampling in the 20's and 30's were primarily elementary static methods. These methods

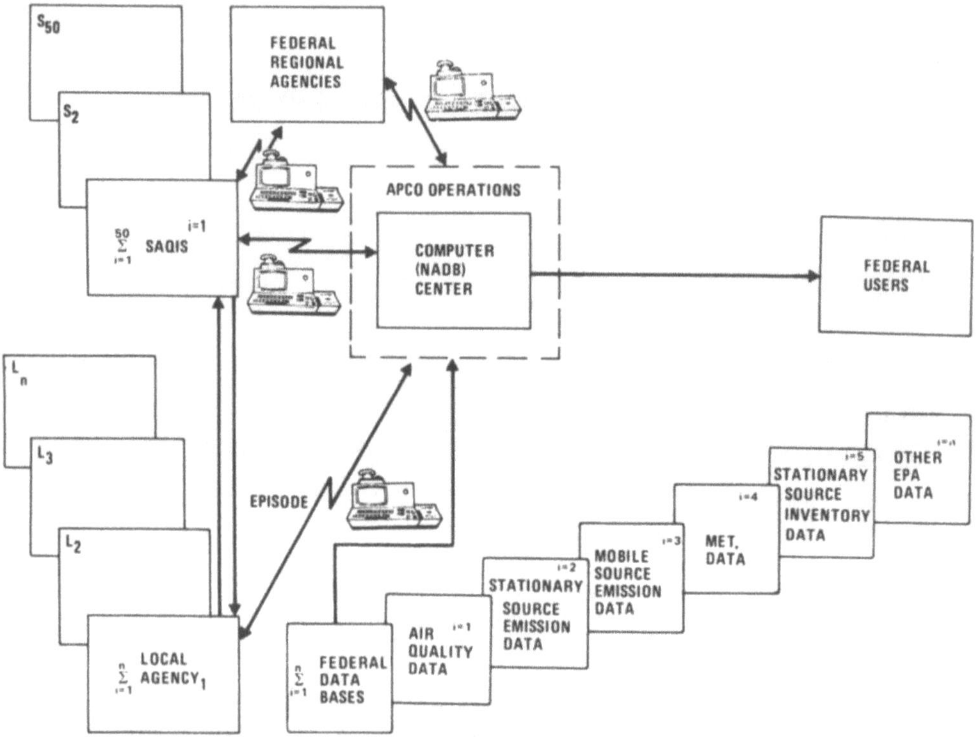

Figure 5. Overview of the National Aerometric Data Information Service

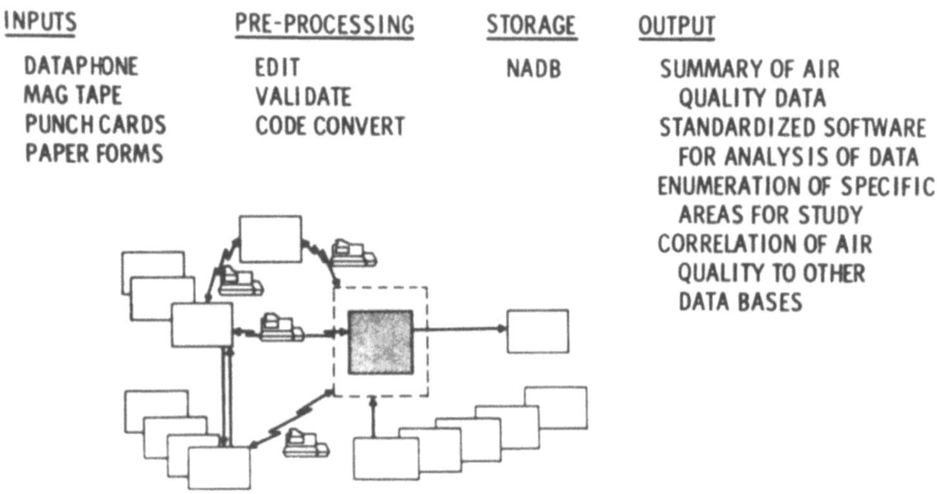

Figure 6. Essential Steps in the Utilization of the National Aerometric Data Bank

indicated the presence of an air pollution problem and gave some general idea of its severity. Physical and physicochemical systems were later developed to quantitatively collect pollutant samples and to allow their subsequent measurement. Initially, traditional wet chemical analyses were carried out in the laboratory by technicians. Automation of these wet chemical methods not only increased accuracy and precision but increased productivity by a factor of approximately 15. New data-recording methods being developed minimize handwork performed by the technician and emphasize the need for adoption of these new techniques for laboratory data handling and validation.

The following new instruments or methods are now at various stages of development or field-testing by EPA.

1) A chemiluminescent method for determining ozone.
2) A flame-luminescent method for measuring sulfur.
3) A triple system for continuously measuring carbon monoxide, sulfur dioxide, and nitrogen dioxide or nitric oxide. The sensors in this system are based upon the fuel-cell principle and time-share the circuitry.
4) An automated gas chromatrograph for the simultaneous determination of carbon monoxide, methane, and total hydrocarbons.
5) An automated gas chromatograph for measuring hydrogen sulfide and certain organic sulfides.
6) A special adaptation of lidar (light detection and ranging) instruments for estimating concentrations of atmospheric particulates.
7) An atomic absorption method for continuously measuring lead.
8) A piezoelectric instrument for continuously measuring suspended particulate concentrations.
9) A specific-ion electrode method for continuously measuring fluorides.
10) A method for determining oxides of nitrogen based on a gas-phase reaction, which produces light by chemiluminescence.

Research is in progress to develop remote sensing devices to be used at ground level, in aircraft, and in earth satellites for area-scanning and profiling. Applications of newer laboratory techniques such as microwave absorption spectrometry, electron or ion microprobe analysis, and neutron activation analysis have recently been reported. In EPA's laboratories emission spectrometry, atomic absorption spectrometry, chromatography, and other techniques are now being computer-interfaced for routine use.

SUMMARY

Through atmospheric surveillance the identity and concentrations of airborne pollutants have been established. Some 50 years ago collecting devices came into use for measurement of airborne particles. The past 25 years have witnessed continued improvement and expansion in the quality and variety of air surveillance methods and instruments. The combined developmental efforts have culminated in a system of instruments capable of collecting and analyzing a variety of pollutants and presenting the data in a machine-readable format.

To meet future needs and responsibilities, Federal, State, and local control agencies must participate in a cooperative and unified system for collecting, storing, retrieving, analyzing, and disseminating aerometric data. This system, designated NADIS, will provide a systematic and timely flow of these data to contributing agencies, research organizations, and the public. The NADIS concept is EPA's flexible and long-range program to provide needed data of high quality upon which decisions will be based to assure the sufficiency and integrity of our environment.

REFERENCES

1. L. A. Chambers, "Air Pollution," Vol. I, by A. C. Stern, Academic Press, New York (1968).
2. J. E. Ives, et. al., "Atmospheric Pollution of American Cities for Years 1931-1933," U. S. Public Health Bulletin No. 224 (1936).
3. H. C. Wohers and G. B. Bell, "Literature Review of Metropolitan Air Pollutant Concentrations" Stanford Research Institute, Menlo Park, Calif. (1956).
4. Second Technical and Administrative Report on Air Pollution Control in Los Angeles County, Air Pollution Control District, County of Los Angeles, Calif. (1950-1951).
5. The Smog Problem in Los Angeles County, Stanford Institute Research Report, Western Oil and Gas Association, Los Angeles, Calif. (1954).
6. First Technical Progress Report, Air Pollution Foundation, Los Angeles, California (1955).
7. Air Pollution Measurements of the National Air Sampling Network, (1953-1957) Public Health Service, Cincinnati, Ohio (1958).
8. Air Pollution Measurements of the National Air Sampling Network (1957-1961) Public Health Service, Cincinnati, Ohio (1962).
9. L. R. Setter, C. E. Zimmer, D. S. Licking and E. C. Tabor, AIHA Journal 22, 192, (1961).
10. E. C. Tabor and C. C. Golden, APCA Journal 15, 7 (1965).
11. G. A. Jutze and E. C. Tabor, APCA Journal 13, 278 (1963).

12. G. J. Nehls, D. H. Fair, J. B. Clements, Env. Sci. Tech. 4, 902 (1970).
13. Inventory of Air Monitoring Equipment Operated by State and Local Agencies, Environmental Protection Agency, APCO, Air Pollution Technical Information Center, Research Triangle Park, North Carolina (1971).
14. G. B. Morgan, G. Ozolins, E. C. Tabor, Science, 170, 289 (1970)
15. Guidelines: Air Quality Surveillance Networks, Environmental Protection Agency, APCO, Air Pollution Technical Information Center, Research Triangle Park, North Carolina (1971).
16. Guidelines for the Acquisition of Validated Air Quality Data, Environmental Protection Agency, APCO, Air Pollution Technical Information Center, Research Triangle Park, North Carolina (1971).
17. Storage and Retrieval of Aerometric Data, User's Manual, Environmental Protection Agency, APCO, Air Pollution Technical Information Center, Research Triangle Park, North Carolina (1971).

IMPORTANCE OF AIR QUALITY MEASUREMENTS TO CRITERIA, STANDARDS AND IMPLEMENTATION PLANS

Delbert S. Barth, Ph.D.

Environmental Protection Agency, Air Pollution Control Office, Bureau of Air Pollution Sciences

INTRODUCTION

The Clean Air Act as amended, P.L. 91-604, states that one of its purposes is "to protect and enhance the quality of the Nation's air resources so as to promote the public health and welfare and the productive capacity of its population." It goes on to define specific authorities granted to the Administrator, Environmental Protection Agency (EPA), that he may utilize to achieve the purposes of the Act. In brief, some of these authorities are:

1. Designation of air quality control regions, issuance of criteria and control-techniques documents, and promulgation of national ambient air quality standards.
2. Promulgation of national standards of performance for new stationary sources.
3. Promulgation of national emission standards for stationary sources of hazardous air pollutants.
4. Promulgation of national emission standards for motor vehicles.
5. Regulation of fuels and fuel additives.
6. Issuance of national aircraft emission standards.
7. Setting of aviation fuel standards*

A brief discussion of each authority will be given, with emphasis placed on the importance of measurements of pollutants to each case. It is important to bear in mind that Federal policy for air pollution control is based on the need to protect the

*This authority is granted to the Administrator, Federal Aviation Administration, but it must be based on recommendations of the Administrator, EPA.

public from the adverse effects of pollutants on health and welfare and to enhance the quality of the total environment.

GENERAL DISCUSSION OF AIR QUALITY

To fully define the quality of air with respect to a specified air pollutant, it is first necessary to define the concentration in ambient air of that pollutant, the length of time air was sampled for that pollutant, and the location(s) to which the observed concentrations are applicable. The description of the location must include height above the ground as well as location on the ground; that is, location must be defined by polar rather than rectilinear coordinates. Furthermore, the sampling device used and its sampling rate must be described as well as the analytical procedure used to quantitate the pollutant of concern.

The fundamental importance of air quality stems from the fact that most effects on human health or welfare are related to pollutant levels at the locations of the receptors. Important exceptions to this statement are effects such as decreased visibility or altered weather or climate, for which a direct relationship between effect and air quality at a single location is clearly not possible. In these cases, the important parameter is integrated air quality through an appropriate layer of the atmosphere.

The concentration of a pollutant at a given receptor is a quantity that usually results from contributions of that pollutant from many different stationary and mobile sources. Thus to achieve an air quality standard, it is necessary to know all the significant sources of the pollutant that contribute to its concentration at a given location. From this knowledge it is then possible to derive a collection of necessary emission standards for the significant sources that will ensure that the air quality standard will not be exceeded.

Even though it is easy to state the principles involved in proceeding from air quality standards to emission standards, it is not easy to apply those principles precisely to special cases for the following reasons:

1. Not all significant sources may be identified.
2. Not all sources may be controllable to the same degree.
3. Meteorological parameters play a dominant role which is not accurately defined.
4. There may be significant non-man-made background sources of the pollutant which are uncontrollable.

The point to be made is that it is not possible with our present state of knowledge to confidently derive a set of emission standards for a given Air Quality Control Region which will assure that the air quality standards will not be exceeded. Thus even after emission standards have been determined, promulgated, and enforced, it will be necessary to continue measuring air quality. Another factor not yet mentioned, which is also difficult to predict precisely, is the future growth of sources of the pollutant of concern.

In general, then, air quality measurements are needed principally for the following purposes:

1. To relate observed effects on health or welfare to pollutant concentration and averaging time (sampling times from such average pollutant concentrations were determined).
2. To determine the control needed to protect the public from adverse effects on health or welfare.
3. To determine efficacy of controls as they are installed.
4. To determine non-man-made background levels.

With the general discusssion of air quality completed, we will now examine some of the authorities contained in the Clean Air Act, as amended, in more detail.

CRITERIA AND CONTROL TECHNIQUES DOCUMENTS

The Administrator, EPA, is directed by P.L. 91-604 to publish criteria and control technique-documents for those air pollutants

"(A) which in his judgment have an adverse effect on public health and welfare;" and

"(B) the presence of which in the ambient air results from numerous or diverse mobile or stationary sources."

Air quality criteria for an air pollutant "shall accurately reflect the latest scientific knowledge useful in indicating the kind and extent of all identifiable effects on public health or welfare which may be expected from the presence of such pollutant in the ambient air, in varying quantities." Information on control techniques "shall include data relating to the technology and costs of emission control. Such information shall include such data as are available on available technology and alternative methods of prevention and control of air pollution." The Act states that "effects on welfare include, but are not limited to, effects on soils, water, crops, vegetation, man-made materials; animals, wildlife, weather, visibility, and climate, damage to and deterioration of property, and hazards to transportation, as well as effects on economic values and on personal comfort and well-being."

For those materials for which criteria and control-technique documents are issued, the Administrator is directed to prescribe national primary and secondary ambient air quality standards, which are defined as follows:

1. "National primary ambient air quality standards shall

be ambient air quality standards the attainment and maintenance of which in the judgment of the Administrator, based on such criteria and allowing an adequate margin of safety, are requisite to protect the public health."

2. "National secondary ambient air quality standards shall specify a level of air quality the attainment and maintenance of which in the judgment of the Administrator, based on such criteria, is requisite to protect the public welfare from any known or anticipated adverse effects associated with the presence of such air pollutant in the ambient air."

To date, criteria and control-technique documents have been issued and National Primary and Secondary Ambient Air Quality Standards have been proposed for sulfur oxides, particulate matter, carbon monoxide, photochemical oxidants, hydrocarbons, and nitrogen oxides. Once these standards have been promulgated, which must be no later than April 30, 1971, the States have 9 months in which to develop and submit an implementation plan designed to achieve the air quality standards within a rather tight time schedule after approval of their plan--3 years after approval for primary standards, and a "reasonable time" for secondary standards.

What is the role of air quality data in all of this? First, air quality data were necessary as a basis for the criteria documents. Second, air quality data are necessary in determining the controls needed to meet the standards. Third, an integral part of the implementation plan must be an adequate surveillance network to ensure that progress toward the standards is measured and documented.

It is planned that all surveillance networks, Federal, State, and local, will report their data in a compatible format so that all data may be stored in one central computer repository and retrieved as required. This will ensure that the appropriate control agencies receive comparable, accurate, and current data. Steps are now being rapidly taken to achieve this goal. As more and more State and local surveillance networks become operational, the Federal Government will reduce its collection of routine enforcement-oriented data. It is expected that the appropriate State and local agencies will take over and continue to operate the existing Federal National Air Sampling Network stations at their existing sites to provide continuity of air quality data at these locations. The Federal Government will increase its collection of research and baseline-oriented data. Examples of such activities include systematic identification and quantitation of new pollutants as well as detailed surveillance on an *ad hoc* basis in the vicinity of selected major point sources in order to validate, and/or provide new data for improving, our existing predictive models. Several Federal mobile air quality measurement laboratories are planned for the future to facilitate this type of operation.

STANDARDS OF PERFORMANCE FOR NEW STATIONARY SOURCES

"The term 'standard of performance' means a standard for emissions of air pollutants which reflects the degree of emission limitation achievable through the application of the best system of emission reduction which (taking into account the cost of achieving such reduction) the Administrator determines has been adequately demonstrated." On a tight time schedule set forth in the Act, the Administrator is required to list and then publish standards for categories of new sources that in his judgment "may contribute significantly to air pollution which causes or contributes to the endangerment of public health or welfare." The measurement of air quality plays no role in this authority of the Federal Government.

The Act stipulates, however, that the Administrator shall prescribe regulations that will establish a procedure, similar to that of the implementation plans for criteria pollutants, under which each State shall submit to the Administrator a plan that establishes emission standards for any existing source in the same category of sources for any air pollutant which is a "non-criteria" pollutant and which has not been listed as a hazardous air pollutant, and to which a standard of performance would apply if the existing source were a new source. Thus all of the uses of air quality data cited under the section of criteria and control-technique documents would then be applicable for the existing source case here.

NATIONAL EMISSION STANDARDS FOR STATIONARY SOURCES OF HAZARDOUS AIR POLLUTANTS

"The term 'hazardous air pollutant' means an air pollutant to which no ambient air quality standard is applicable and which in the judgment of the Administrator may cause, or contribute to, an increase in mortality or an increase in serious irreversible, or incapacitating reversible, illness." The Act requires the Administrator, on a tight time schedule, to publish a list and then, subsequently, to promulgate national emission standards for those air pollutants deemed hazardous. The emission standards will be applicable to both new and existing stationary sources. In general, standards will be set by defining internally "reference air quality standards" at the property line of applicable sources and then using dispersion models to back-calculate to those allowable emissions that will ensure that the reference air quality standards will not be exceeded. The calculated allowable emissions would then serve as a basis for the national emission standards.

Thus air quality data are needed to develop, improve, and validate prediction models linking sources to air quality as well as to verify subsequently that the controls installed achieve the desired results.

NATIONAL EMISSION STANDARDS FOR MOTOR VEHICLES AND AIRCRAFT AND REGULATIORY AUTHORITY FOR FUELS AND FUEL ADDITIVES TO INCLUDE AVIATION FUELS

For all of these authorities it is envisioned that control needs will be based on allowable levels of significant pollutants to assure protection of health and welfare. Thus, as noted in the last section, air quality data are needed for development, improvement, and validation of prediction models linking sources to air quality and for subsequent verification that the desired results have been achieved.

SUMMARY AND CONCLUSIONS

Air quality measurements are mandatory for the proper application of many of the authorities contained in the Clean Air Act as amended. Thus, in order to get the maximum return for Federal, State, and local resources devoted to air quality measurements, it will be essential in the future to develop improved measurement techniques; to optimize monitoring system design; to develop a common system of calibration, quality assurance, data validation, and reporting; and to strive for integrated (with a minimum of overlapping) Federal, State, and local air quality surveillance networks. We are currently working hard to achieve these ultimate goals.

AEROMETRIC DATA: NEEDS AND NETWORKS

Dr. T. R. Mongan (1) & E. L. Keitz (2)

(1) Sydney Area Transport Study, Sydney, Australia

(2) The MITRE Corporation, McLean, Virginia 22101

The primary business of air pollution control agencies is the prevention, control, and abatement of air pollution. Air pollution is a local problem whose nature depends on the emission sources, meteorology, and topography of the affected area. It is a nation-wide, as opposed to a national, problem, Furthermore, since control must ultimately be accomplished in the affected area, the Clean Air Act states that "the prevention and control of air pollution at its source is the primary responsibility of state and local governments."

The control of air pollution is more a legal, social, and economic problem than a technical problem. Nevertheless, there are important technical aspects to the control problem and some of these aspects cannot be accomplished without valid air quality data. Thus, although the accumulation of valid aerometric data is definitely of secondary importance in air pollution control work, it is necessary. The optimum air monitoring network for a control agency is that network which supplies the aerometric data necessary to support the agency's prevention, control, and abatement activities, while consuming the minimum amount of the agency's financial and manpower resources. This is not to say that aerometric networks providing data for scientific or research purposes are not useful and valuable, but simply to stress the point that a control agency's main objective should be to control air pollution, not merely to study it.

This paper discusses the effective uses of aerometric data for the prevention, control, and abatement of air pollution and the data needs arising from these uses. Our views are based on an extensive study of air monitoring networks, including on-site inspection of many aerometric data collection networks around the United States, discussions with the manufacturers of air monitoring and aerometric data handling systems, careful review of many proposals for air monitoring networks, and extensive experience handling and studying the data produced by air monitoring networks across the country. We present an approach which will enable a state or local control agency to design or review an air monitoring network, insuring that the network will provide valid data which will be used for the prevention, control and abatement of air pollution with minimum drain on the agency's financial and manpower resources.

USES OF AEROMETRIC DATA

The development or review of any network for the accumulation of aerometric data must start with a clear and complete description of the uses which will be made of the data. Uses of aerometric data generally fall into three categories: operational, legal and public information.

The operational uses of air quality data are those intended to support or trigger effective action to protect or improve air quality within an agency's jurisdiction. These uses of aerometric data are by far the most important uses, because they may have a direct beneficial effect on air quality. Three examples of such operational uses of data follow.

If data, proving that ambient pollutant concentrations in the vicinity of a source exceed standards, are necessary to the successful conclusion of a legal action which will ensure the abatement of pollution from the source, then no effort should be spared in the attempt to accumulate the necessary data.

It is far more desirable to prevent the development of an air pollution problem than to try to control the problem once it has developed. Consequently, air pollution control agencies should make every effort to establish strong ties with planning and zoning agencies within their jurisdiction. When development is contemplated for an area, the pollution control agency can do some sampling in the area. This can be a very effective use of aerometric data, and considerable efforts to ascertain the data required by a planning agency and subsequently to obtain the necessary data at minimum cost are certainly justified.

If a control agency is legally required to show that air quality has deteriorated past a certain point before declaring an alert or instituting administrative control actions which will be <u>effective</u> in controlling or abating air pollution within the agency's jurisdiction, this is certainly an important use of data. However, unless an agency is legally required to refrain from such actions, it is far better to take such action on the basis of meteorological forecasts, thus preventing the occurrence of severe air pollution problems rather than trying to control them once they exist.

The second type of aerometric data usage is the provision of air quality data which an agency is required by law to collect. This use is important because, by law, an agency must gather such data. However, since the gathering of such data does not of itself directly influence the air quality in an agency's jurisdiction, a careful and complete description of the data which the agency is legally required to produce should be compiled and then the agency should attempt to provide these data at minimum expenditure of manpower and funds. According to Section 110 of the Clean Air Act, each state must adopt, for each pollutant for which a national ambient air quality standard is promulgated, "a plan which provides for implementation, maintenance, and enforcement" of the standard. However, federal law contains no direct requirements for data from pollution control agencies below the state level, nor does it specify the number of sites at which the states must measure pollutants.

In almost every situation, for public information purposes, very little data is required and such data can be easily obtained. Providing data as a public information service is useful for generating public pressure for prevention, control, and abatement legislation and encouraging public cooperation and support for control efforts. It is important that data used for public information be presented in brief, concise, and readily understandable fashion. The public is not interested in masses of confusing data.

DOCUMENTING DATA USES

The first step in designing or reviewing an aerometric data collection system is to carefully and completely list and document the uses which will be made of aerometric data. This list of uses must be a detailed list. It is inadequate to simply write down "operational use - obtain background data" or "public information use - release to press."

Once such a clear and detailed statement of the uses which an agency makes of its aerometric data is available, it will be easy to determine whether the cost of an agency's aerometric data collection effort is justified by the uses made of the data. All too often in the past, a great deal of money had been spent to accumulate air quality data which was never used. This must be stopped.

There are at least two serious misconceptions which have arisen concerning the use of aerometric data in air pollution control work which must be discussed before proceeding further. The first is the misconception that "we must gather a great deal of aerometric data so we can 'understand' air pollution, for we cannot control air pollution until we 'understand' it." The logical fallacy in this statement can be seen by imagining the consequences of eliminating all sources of air pollution in an area. Air quality could be expected to show a marked improvement, regardless of the fact that the control agency had no understanding whatever of atmospheric chemistry, diffusion, etc. The operational fallacy of this first misconception can be seen by observing, for example, that a great deal more could be done for the air quality in a city by preventing the combustion of high sulfur fuel than by measuring or understanding the distribution of sulfur dioxide in the atmosphere. Unfortunately, this misconception has led some control personnel to forget their primary responsibility of prevention, control and abatement of air pollution and to try to take over the scientist's role, diverting an unnecessarily large amount of time and resources from the primary task of control into measurement of pollutants, attempts at "characterizing the air mass over the area," and establishment of elaborate data manipulation and analysis programs.

The second common misconception is "we should gather all the aerometric data we can, because you never can tell what it will be used for." This, and the somewhat more sophisticated misconception we mentioned first, seem to be the only justification for some aerometric data collection networks. In opposition, we stress again the obvious fact that the business of an air pollution control agency is control, not the accumulation of aerometric data. Unless the data accumulated by a data collection system is being used, the money spent on accumulating that data has been wasted. The additional data might be of interest for scientific purposes, but if it will not be used for control work by the agency, it has no value to the agency.

Air pollution control agencies should only accumulate data which will be used on control work. The aerometric data accumulated should be accurate, reliable, and validated data. This is especially important if the data will be challenged in court. It is, therefore, far better to have a few carefully located, well maintained, and calibrated sensors providing data to a well designed data processing system than a large number of poorly located, inadequately maintained sensors providing data to a substandard data processing system. Bad data can be worse than no data at all.

NEEDS FOR DATA

When a control agency has prepared a complete and detailed list of the uses for aerometric data gathered by the agency's data collection system, the next step in planning or reviewing a data gathering system is to ascertain the agency's actual needs for aerometric data. To do this, the agency should prepare a detailed specification of the minimum amount of data needed for each use of aerometric data which has been specified. Thus, for each use to be made of aerometric data, it should be determined:

1) What pollutants must be monitored?
2) Where must each pollutant be monitored?
3) How often must measurements be taken?
4) What averaging time should be employed?
5) What maximum delay in reporting readings to the agency headquarters is tolerable?
6) What processing, analysis, and preparation of the raw data must be performed?

Since an effective network should be carefully tailored to meet an agency's needs for data, the greatest caution and precision should be exercised in preparing a careful and detailed list specifying the agency's minimum need for data. Inclusion of any data above the minimum needed for the uses which the agency has specified is unnecessary and can greatly increase the cost of the network which must supply the data needs. Any well-designed aerometric data collection system can be readily augmented, and effective air monitoring systems can and do grow in response to growth in the needs for the data which they produce. So, there is nothing to be lost by carefully specifying the minimum data required by the agency for the various uses to be made of aerometric data. If an agency is not careful to specify their minimum data needs, much more money and effort can be wasted than that required to deploy unneeded instruments at unnecessary stations. For example, the specification of an unnecessarily short maximum allowable delay in reporting readings to agency headquarters might incline an agency toward deploying an elaborate and expensive telemetry system which is not really needed. It should be noted that some agencies may be able to satisfy all or part of their data needs by simply taking data provided by a network operated by another agency. This should be done whenever possible, for there is then no reason or justification for duplicating data which is already available.

Among operational uses, the use of source-oriented data in legal actions requires the monitoring of those pollutants which the agency is trying to control, based on known or suspected emissions from the source. Similarly, the use of data for planning studies requires the determination of the existing levels of pollutants which might be expected to become a problem if the planned sources were to be established. Legal uses of data require the monitoring of pollutants for which standards have been set. Only those pollutants which represent a problem in the agency's jurisdiction because of known or suspected emissions should be monitored for public information uses.

For most operational uses of data, the location of monitors is rather obvious; source-oriented monitoring must be done near the source, background monitoring for planning purposes must be accomplished in the area where development is projected and, if necessary to trigger alerts, surveillance monitoring should be performed in the area where pollutant levels are expected to reach high levels early in an episode. For legal uses, it may well be unnecessary to monitor at more than one site within an agency's jurisdiction. Moreover, careful consideration should be given to the use of movable (as opposed to truly mobile) monitoring stations. Obtaining data for public information uses should, on the other hand, always be done at fixed locations so the data released to the public has some continuity. Furthermore, such monitoring should be done at a site where the measurements have meaning to the public, providing some indication of the maximum pollutant concentrations to which the population is being exposed. Consequently, if a site for public information monitoring is being chosen, it should be located in the area where pollutant levels are highest, chosen from among those areas in the political subdivision frequented by people.

Measurements should be taken only as often as necessary. Particularly in conjunction with legal uses of the data, careful consideration should be given to intermittent sampling. Intermittent sampling should be performed more frequently in highly polluted areas than in areas with lower pollutant concentrations. If an agency is severely limited in manpower and resources, it might be possible for them to meet their legal needs for data by the use of intermittent sampling using a movable station placed at one site at a time. For public information uses, it might be acceptable to sample only three days a week, while for operational uses, sampling should only be done when it is really necessary.

The averaging time employed should always be the averaging time mentioned in the standards applicable to the pollutant measured. If there are no standards established for a pollutant, the averaging time is to some extent governed by the choice of pollutant monitoring equipment. However, if the available instruments will allow a choice of averaging times without added cost and furthermore, if standards have not been established, pollutants should be monitored with the least expensive effective method available, accepting whatever constraints this may put on averaging time. When pollutants for which legal standards have not been set are being monitored, the choice of averaging time should be influenced by the following reasoning. As a general rule, the averaging time for measurements on a physical system should be somewhat less than the characteristic time for significant changes in the physical property being measured. Significant changes in ambient air quality can occur in times on the order of a few hours. However, since short-term fluctuations in pollutant levels do occur, average measurements are preferable to intermittent measurements. Thus, one-hour averaging is quite adequate for recording ambient air quality. At present, short-term local fluctuations are only of interest to research scientists. Thus, averaging times of less than one hour are not needed for ambient air monitoring unless shorter averaging times are demanded by standards. Finally, the Air Pollution Control Office (APCO) of the Environmental Protection Agency recommends that 24-hour averages should be taken over the period midnight to midnight to allow comparison with standard meteorological summaries.

The allowable delay in reporting data to agency headquarters is determined by the uses of the data. Legal uses usually have the longest allowable delay, because there is rarely a deadline on data presentation. Allowable delay in receipt of data for public information uses is usually around 12 hours to one day. Generally, reducing the delay time in receipt of data tends to greatly increase costs by requiring expensive telemetry, trips to monitoring sites or manned stations, or telephone communication of data. When we turn to operational uses for data, the only justification for specifying a maximum allowable delay in reporting data to agency headquarters of T hours is that effective action can be initiated in a time on the order of T hours which will have a substantial effect in preventing further degradation of air quality within the agency's jurisdiction. Obviously, there is no need for short reporting times when data is being accumulated for lawsuits or for land use studies. In fact, during our extensive studies of air monitoring networks, we have encountered no instance in which there was a clearcut need for a reporting time less than a couple of hours. If declaring an episode or alert does not lead to a cessation or reduction of emissions, (regardless of whether or not an alert is called on the basis of air quality measurements) the provision of a system (such as telemetry) which allows short data reporting times is useless, since it merely allows the agency to watch the air get dirty in "real-time." We have, as yet, seen no justification for telemetry systems in air monitoring on the basis of their ability to provide data with short delays. However, at such time as reliable pollutant sensors which can be left unattended in the field for long periods of time become available, telemetry might become desirable because of the savings of manpower expended on data collection trips which might be reduced. However, this situation is far from the reality of today's air monitoring capabilities. Currently, monitoring sites must either be continuously manned or visited so frequently to pick up high-volume samples and perform sensor maintenance and calibration that virtually all data needs can be met by hand carrying data to agency headquarters or by making a telephone call.

All effective uses of air quality data depend on the availability of valid data. Consequently, some sort of a data processing system is a necessary part of every effective aerometric data collection system. Essentially, such a data processing system must accept raw data from the sensors, validate that data, and prepare the data in a form suitable for use. The basics of the necessary data processing system are the same whether all the work is performed by a single man or the system involves several people and some electronic computers. Validation must always be keyed to inspection of the raw data by an air pollution professional familiar with both the sensors and data recording and transmission systems used to gather the data and the air pollution problem within the agency's jurisdiction including meteorology, topography, and source emission characteristics.

In all cases, the necessary averages specified in the standards must be calculated and the data must be prepared in a form suitable for use. Generally, both arithmetic and geometric means and standard deviations should be calculated and high pollutant concentrations and their times of occurrence noted. Some agencies may be required to prepare frequency distributions in order to demonstrate compliance with state or local standards. In particular, a careful identification of the location of the monitoring site and the monitoring equipment used should always accompany any data released by an agency, in order to avoid misinterpretation of the data. Parenthetically, it is clear to anyone involved in air pollution control work that data validation, analysis, and preparation of data for use can consume a great many valuable man-hours. In fact, it is a good rule of thumb that a minimal computerized data processing system should be considered whenever data processing begins to consume more than one man-week of professional time per month. Note that if air pollution control efforts continue and air quality improves, there may be some reduction in the need for aerometric data for control purposes. This is especially true of the operational uses wherein data is needed from source-oriented monitoring for enforcement purposes.

It may appear that the detailed specification of an agency's uses of aerometric data and the careful listing of the resulting data needs which we recommend is a great deal of trouble, especially for a small agency. However, our approach can save a great deal of money and manpower, especially for a small agency. This is because air monitoring consumes a large part of the resources of a small agency and any unnecessary expenditure of money or manpower on aerometric data collection can have a particularly detrimental effect on a small agency's control efforts. Our approach of detailing the uses of aerometric data, as well as the needs for data springing from these uses, should always be used in reviewing an air monitoring network because:

- A comparison of the uses made of aerometric data with the cost of the air monitoring effort will make it very clear whether the overall data collection program is cost-effective and worthwhile; and

- A comparison of the agency's data needs, arising from the uses made of the data, with the data actually provided by the agency's aerometric data collection system will demonstrate whether or not the system is well designed.

DESIGNING NETWORKS ON THE STATE AND LOCAL LEVEL

Once a clear and complete listing of the uses which an agency will make of aerometric data, accompanied by a detailed specification of the resulting needs for data, has been prepared, the most effective air monitoring system for a control agency can almost be said to design itself. An agency's aerometric data collection system should be designed by choosing the least expensive system which will meet the agency's minimum data needs, providing data for effective use in the prevention, control, and abatement of air pollution.

We shall not attempt to provide hard and fast rules for network design, definite numbers of monitoring sites which should be employed, or exact specifications for data processing systems, because the entire thrust of our argument is that aerometric data collection networks must be carefully tailored in each individual case to meet a particular agency's special needs for data. This approach to network design is the only valid approach in a field like air pollution where the uses of data, needs for data, and the very nature of the air pollution problem varies widely from area to area owing to the peculiarities of meteorology, topography, source characteristics, and legal and administrative situations.

The basis of our design approach is to design a network which will provide aerometric data which can be effectively used to control air pollution at minimum cost to the agency operating the network. This is done by choosing the lowest cost design options, including purchase, maintenance, and operation costs, which will provide data meeting the agency's carefully identified list of needs for data. In addition, in order to remain within the limits imposed by the money and manpower available to an agency for air monitoring efforts, it may be necessary to establish some order of priority among the data needs. This priority will usually become clear from the uses to be made of the data, but in some cases the priority ordering can be sharpened or altered by the available options in network design.

Every aerometric data collection system can be divided into three subsystems:

- Sensor Subsystem;
- Data Recording and Transmission Subsystem;
- Data Validation, Analysis, Storage, and Retrieval Subsystem.

Although we shall consider system design by treating each subsystem separately, it must be held in mind that it is total system cost which must be minimized, not subsystem cost.

Sensor Subsystem

Where must monitoring be done? This must be determined from the needs for data. If the agency has been unable to specify the exact location for monitoring in their statement of data needs, then the location of the monitoring sites must be specified at this time. For example, if monitoring is to be done at a site in the zone of highest pollution in the agency's jurisdiction, this must be determined from meteorological considerations, a rough source inventory (which should be the first order of business at any new control agency), and diffusion calculations (if necessary), to establish the general area where monitoring is to take place. In every case, the exact location of a monitoring site must be pinpointed on the basis of accessibility, availability of space, and the needs for electric power and security. These latter considerations, of course, can severely constrain the choice of a monitoring site. If a choice must be made between sites, choices should be made on the basis of the following list of priorities:

- Sites for source-oriented monitoring for enforcement purposes. These data can be used in lawsuits or they can be compared with standards and published in the press, which may lead to the development of control legislation if none exists.

- A site in the zone of highest pollutant concentration in the agency's jurisdiction. This site is important because data from this site can warn of developing air pollution problems in the agency's jurisdiction. Data from this site are also useful for meeting legal needs since, if concentrations in the zone wherein pollutant concentrations are expected to be highest are within the standards, it is likely that the rest of the agency's jurisdiction will be in compliance with standards. Data from this site is also useful for public information needs. Since the health effects of air pollution are a primary cause of citizen concern with air pollution, citizens may want to know the maximum concentrations of air pollutants to which they may be exposed. Monitoring in the zone of highest pollutant concentration also provides a uniform basis for comparison of the air quality in different air sheds in different jurisdictions.

- Sites where background studies are being performed for areas where industrial development is imminent. These sites are important because the resulting data can be used in the prevention of future air pollution problems.

- Sites in areas of high population density. These sites are important for public information uses. Again, since health effects are one of the primary causes of citizen concern about air pollution, citizens might want to know about the levels of pollutant concentration in the areas where they live and work.

- Sites where background studies will be performed in areas where further development is not imminent.

One fundamental part of the data generation phase of air monitoring has been widely neglected. That is the provision of site identification data. It is well known among professionals in air pollution control that data cannot be properly interpreted unless the conditions under which the data were gathered are known. All too frequently, detailed information on monitoring sites, instruments, and the methods employed are unavailable even to the agency which has accumulated the data. This omission seriously reduces the usefulness of the agency's data.

<u>At each monitoring site, what pollutants are to be monitored?</u> Again, the answers to this question should be clear from the agency's list of data needs. If it is necessary to set priorities because of limitations of resources, the following ordering of priorities should be employed:

- Those pollutants which represent a problem in the agency's jurisdiction. This can be determined from known or suspected emissions in the agency's jurisdiction or from prior monitoring efforts.

- Pollutants for which legal standards have been set.

- Pollutants expected to become a hazard to the public health and welfare at some time in the near future.

We repeat, do not monitor the same pollutants at each site unless there is a clear-cut need to do so. Pollutants which are being sampled strictly to meet legal needs or because they might become a problem in the future should be monitored at only one site, if possible, located where pollutant concentrations are expected to be highest. Movable and/or intermittent monitoring should be carefully considered for these pollutants.

<u>What types of instruments should be employed for air monitoring?</u> Some guidelines are available here. For the following pollutants, the choice of instruments seems obvious. For total suspended particulates, high-volume samplers should be used. For carbon monoxide, non-dispersive infrared instruments should be employed. For hydrocarbons, flame ionization monitors should be used. These instruments are specified in the recently proposed federal standards.

For sulfur dioxide or sulfur oxides, oxidants, and nitrogen oxides, the choice is not quite so clear. The availability of good methods is quite limited. Generally, wet chemistry methods are a great deal of trouble and should be avoided if possible. However, for nitrogen oxides, there seems to be no alternative at present. Again, for oxidants, the wet chemical methods may be the best alternative. For sulfur oxides, flame photometry would be superior if there were no other sulfur bearing species in the ambient air surrounding the monitoring station or if species other than sulfur oxides could be scrubbed out of the sample gas stream. In any case, whatever method is used must be referenced to the wet chemical methods prescribed in the standards for nitrogen oxides, sulfur oxides, and oxidants. For pollutants for which standards have not been set, wet chemical methods should be avoided wherever and whenever possible. Whenever possible, an agency should choose between instruments on the basis of reliability. A cheaper but less reliable instrument, over the long run, will be one of the most expensive purchases an agency can ever make.

What kind of maintenance must be performed and on what schedule? Except for the simplest instruments such as sulfation plates and high-volume samplers, air monitoring instruments (which, after all, are measuring very low concentrations of pollutant substances) are very delicate instruments and should be frequently maintained and calibrated if valid data are to be maintained. A maintenance plan, i.e., a routine schedule of maintenance to be performed by a trained technician, is a must. Instruments should never be purchased until the personnel of the control agency purchasing the instruments can maintain and calibrate the instruments. Manufacturers are almost invariably optimistic about the necessary frequency of maintenance for their instruments. The best source of information on the required frequency of maintenance for an instrument is people working in the air pollution control field who have had prior experience with the instrument. It is important to keep careful maintenance records. Such records could have an important impact in case of court action contesting the agency's data.

What kind of calibration must be performed and how often? This should be determined by taking note of the manufacturer's recommendation and, more important, the recommendations of people involved in air pollution control work having prior experience using the instrument. The only truly valid form of calibration is dynamic calibration, wherein the air monitoring instrument is fed a sample gas stream containing a fixed concentration of the pollutant gas of interest in the proper range. Almost all instruments need frequent calibration by skilled technicians. Manufacturers are almost always optimistic about the frequency of calibration required for their instruments. An instrument should not be purchased until the agency purchasing the instrument has the facilities and the personnel to properly calibrate that instrument. Careful calibration records must be kept, noting the time of calibration, the zero drift, and the error on the span gas. These records may be especially important if a control agency's data are challenged in a lawsuit. These records may also help to pinpoint instrument malfunctions. Remote "electronic" calibration which merely checks to see if the instrument is functioning is inadequate. The same holds for so-called "static" calibration in which, for example, the reagent in an instrument whose readings are based on the conductivity of a solution which has absorbed pollutants is replaced by a substance of known conductivity. This approach fails to take into account the collection efficiency of the air pollution monitoring instrument.

The amount of laboratory support required can be seriously underestimated. A good instrument technician and facilities for maintaining and calibrating the instruments used by an agency are a must. It is not wise to rely on outside help. Effective use of data in lawsuits may require that instruments be maintained and calibrated in an agency's own laboratory.

Data Recording and Transmission Subsystem

What type of data recording system must be used? Whenever possible, data should be recorded in machine-readable form. If this recording in machine-readable form is not accomplished at the monitoring site, it should be accomplished as soon as possible back at the agency headquarters. The problems caused by the necessity to read data from strip chart recorders are well known to everybody involved in air pollution control work. However in some cases strip chart recorders can be useful for maintenance and calibration work, and they may be useful for providing a permanent record of data for legal and enforcement work.

What type of data transmission system must be employed? The furor over telemetry in air pollution control circles has caused data transmission systems to receive far more attention than they really deserve. Remember, if an agency can take no effective actions based on air quality data it receives, it does not matter when the data accumulated at a distant location arrives at the agency's headquarters. The only criterion in selecting data transmission systems is the one of minimum cost. This is the case if an agency takes action based only on meteorological predictions. However, if, for an example, an agency can take effective action based on air quality data in a matter of hours, then a reporting time delay on the order of an hour is justified, and one must choose the minimum cost system which will provide one-hour reporting times. Likewise, if an agency could take effective action based on air quality data which will have a significant beneficial effect on air quality in their jurisdiction in a matter of minutes (and we have not yet identified an agency which can do this), then a data reporting time on the order of minutes would be required and attempts to obtain a data transmission system which could provide such reporting times would be justified. Remember that if action is taken on the basis of meteorological forecasts (which, at present, is usually a good approach) there is really no need for air quality data with short reporting times.

Most air monitoring instruments must be maintained and calibrated so frequently that it is usually sufficient to pick up data from on-site recorders during maintenance and calibration visits or when picking up samples from high-volume samplers or gas bubblers located at the site. If monitoring stations must be manned, rapid data transmission can occur by telephone when necessary. When and only when, instruments become available which can be left in the field unattended for long periods and there is a need for data with reporting times substantially shorter than the intervals between maintenance and calibration visits should telemetry be employed, for then manpower will be saved. This is not yet the case. During our study of networks, we have not encountered a definite need based on effective use of the data for aerometric data with short reporting delays. In the few borderline cases we have seen, data with short delay times could easily be supplied by telephone from the monitoring site in those instances when it was really needed. A telemetry system cannot be justified by its ability to handle large amounts of data if large amounts of data are not needed. Unfortunately, there are instances in which far more data than is needed and used is being transmitted over telemetered systems. When the time does come to deploy telemetered data transmission systems, the systems deployed should be as simple, cheap, and uncomplicated as possible.

Data Validation, Analysis, Storage, and Retrieval Subsystems

A detailed data processing plan covering validation, analysis, storage, and retrieval of aerometric data is an essential part of any acceptable aerometric network design because the raw data produced by the sensor and data recording and transmission subsystems are rarely usable as is. In fact, the processing of the raw data is almost as important as the actual sampling. Any discussion of aerometric data processing invariably brings up the subject of data formats. The importance of data formats has been both overestimated and underestimated. It has been overestimated because of the choice of a format simply involves choosing a systematic way of listing data and need not constrain an agency in any way. The importance of formats has been underestimated because the availability of data in a clearly defined format, with a complete description of that format available to all users and potential users of that data, is the very basis of an effective data processing system. All too often, data have been recorded in a format which is complicated, confused, incomplete, and insufficiently documented. When such data leave the immediate care of the person responsible for recording them, their utility is effectively destroyed. Thus, a good format for aerometric data need only be:

- Clear and understandable,
- Well documented, and
- Suitable for interchanging data between various agencies.

The standardized SAROAD format recommended by the Air Pollution Control Office (APCO) fulfills all these requirements. Any data recorded in the SAROAD format will be immediately accessible to all users and exchange between agencies will be facilitated.

How will data validation be performed? Data validation always involves the checking of raw aerometric data provided by an agency's air monitoring system by an air pollution control professional with an intimate knowledge of the sensors and data recording and transmission subsystems used by the agency, as well as the peculiarities of the air pollution problem in the area. Essentially, the air pollution control professional must check the raw data once it has been recorded to:

- determine if it has been properly recorded;
- make corrections for known instrument, recording, or transmission system errors or malfunctions;
- check abnormal values and rapid changes of parameters to see if these are plausible physical effects or changes that might be ascribed to as yet unsuspected sensor, data recording, or transmission subsystem malfunctions.

The important question in validation is whether or not a computer should be used to aid in the validation process. Relying on our rough rule of thumb, we would say that a minimal computerized data processing system should be considered if the agency is spending more than about one man-week per month on data processing.

What kind of data analysis should be performed? Once aerometric data has been validated, data analysis can be performed. Data analysis should usually include calculation of arithmetic and geometric means and standard deviations and noting the peak pollutant concentrations and the times and dates of occurrence. More elaborate analyses are time-consuming even if they are done by computer because time must be spent in handling and reviewing them. Consequently, unless there is a definite need arising from a clear-cut use for more elaborate analyses, they should not be performed.

What data storage and retrieval system should be employed? Data storage must insure the records against physical deterioration. Storage procedures must also be systematic and secure. Careful and complete records of the location and content of stored records must be made to avoid the possibility of their being misplaced or lost. Systematic storage guarantees that past data can be quickly and easily retrieved. Secure storage will prevent the premature release of data to be used in lawsuits and enforcement actions. Data retrieval involves the establishment of procedures whereby data can be readily located, retrieved, used, and returned to its proper place in storage. The effective design of such a system simply involves choosing the least expensive system which will accomplish the necessary aims.

REVIEW OF NETWORK DESIGN APPROACH

The approach which we propose for designing or reviewing of air monitoring data accumulation systems for state or local air pollution control agencies is as follows:

1. Make a careful and complete listing of the uses which the control agency will make of the aerometric data gathered by the system.

2. Make a careful and detailed list of the needs for data generated by the agency's uses of the aerometric data.

3. Follow the step-by-step approach set forth above to design an aerometric data collection network carefully tailored to the needs of the air pollution control agency, always keeping in mind the necessity of minimizing the funds and manpower expended on aerometric data collection.

Once this procedure has been followed, there should appear an almost clear-cut path to the design of an efficient, cost-effective air monitoring system. If this approach is followed, an agency will then be able to devote the maximum amount of manpower and funds toward the prevention, control, and abatement of air pollution.

A PROGRAM OF COMMUNITY HEALTH AND ENVIRONMENTAL SURVEILLANCE STUDIES (CHESS)

C.M. Shy, J.F. Finklea, D.C. Calafiore, F. Benson,
W. Nelson, V.A. Newill
Community Research Branch, DER, BAPS, ORM, EPA

411 W. Chapel Hill St., Durham, N.C. 27701

CHESS is an acronym for Community Health and Environmental Surveillance Studies. In this paper, we describe the objectives, structure and operating methods of the CHESS program, which is conducted by the Community Research Branch, Division of Effects Research, Office of Research and Monitoring (Environmental Protection Agency), in cooperation with local public health agencies. Our basic intent is to relate community health to changing environmental quality. CHESS consists of a continuing series of epidemiologic studies conducted in selected communities representing an exposure gradient for the most common air pollutants. The program involves monitoring of environmental quality with simultaneous surveillance of health indicators known to be sensitive to variations in environmental quality. The purpose of the CHESS program is to evaluate environmental standards, quantitate pollutant burdens, and document the health benefit of air pollution control.

The CHESS program has three key elements as shown in Table 1: area sets, sensitive health indicators, and environmental monitoring.

AREA SETS

An area set consists of a group of communities, usually three or four, selected to represent an exposure gradient for a pollutant, but similar with respect to climate and socioeconomic traits. Each community in an area set is a defined middle class residential segment of a city containing three or four elementary schools of 500 to 1000 children per school and often a secondary school. In the Southeast, duplicate communities were chosen to represent black and white segments of the population.

TABLE 1

ELEMENTS OF THE CHESS PROGRAM

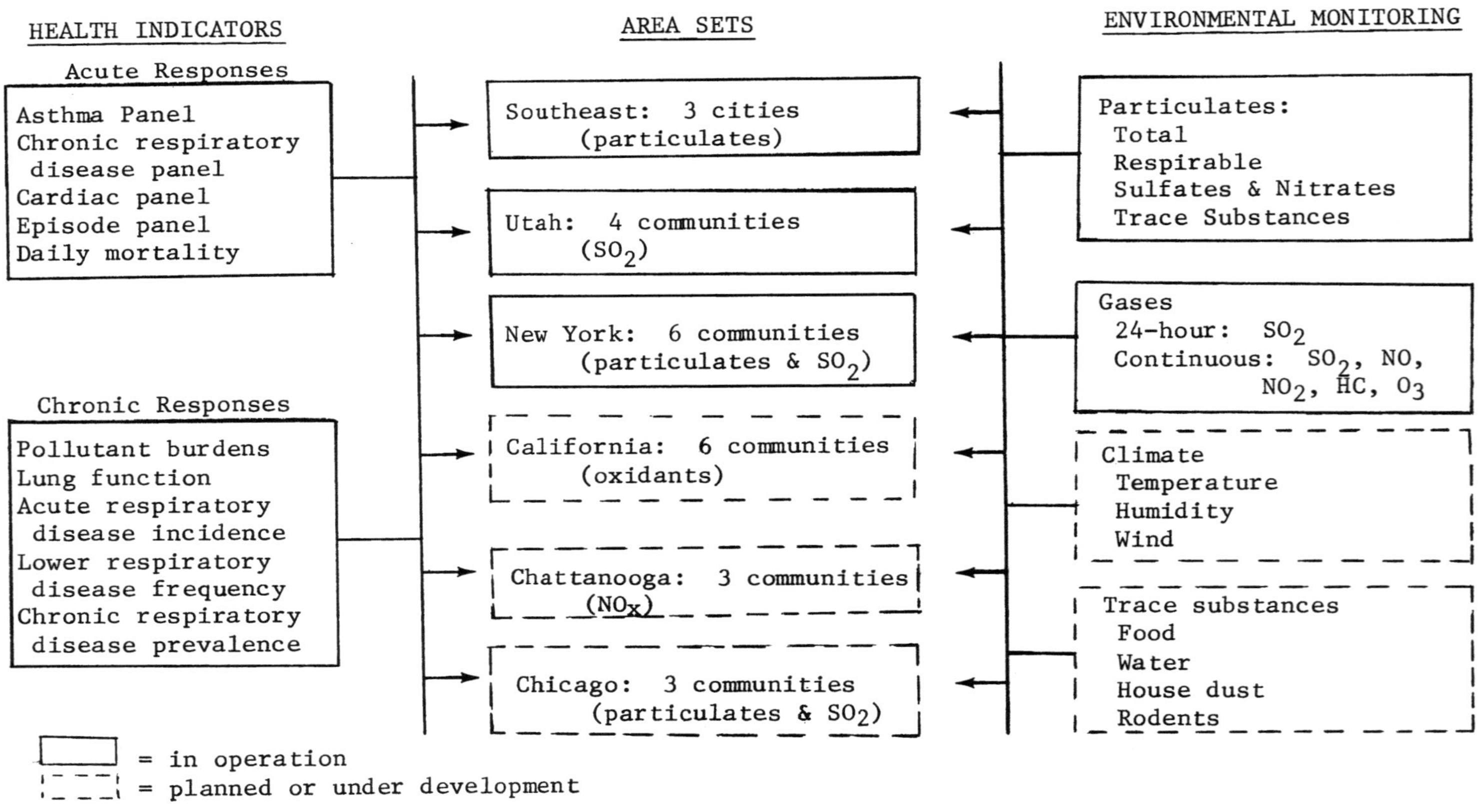

We intend to operate the CHESS program in each area set for a minimum of five years. In the first years, communities within the same set are expected to show a gradient in pollution exposure. Over time, this gradient should diminish with control of air pollution levels. By measuring sensitive health indicators through this period of change, we hope to document and quantitate the health benefits of air pollution control.

HEALTH INDICATORS

In the CHESS program, we measure pollutant effects over a broad spectrum of human responses, including: no demonstrable effect, increase in body burden, altered physiology of uncertain significance, physiologic sentinels of disease, acute and chronic disease and death. We may conveniently divide CHESS health indicators into acute and chronic responses, as shown in Table 2. For the most part, we observe acute responses by systematically following pre-enrolled panels of subjects and comparing response frequency against daily variations in pollutant levels. Observations in communities having consistently low exposure allow us to quantitate the simultaneous effects of environmental covariates such as season of year, temperature and other climatologic changes. We relate acute response frequencies to pollutant concentrations measured by continuous or 24-hour integrated samplers.

We expect the chronic responses listed in Table 2 to reflect effects of exposure to long term average pollutant levels or of repeated exposure to short term peak levels. We do not hope to discriminate between effects of long term exposure and repeated short term peak levels by means of community studies with uncontrolled exposures.

In addition to the pollutant exposures monitored in the CHESS program, we obtain information on other community determinants of response frequency. These determinants or "covariates", along with environmental pollution, co-determine the quantitative level of the health indicator in the community. Covariates include (1) demographic variables such as age, sex, race, socioeconomic level, family size, (2) environmental variables such as occupational dust exposure, geographical migration, intake of pollutants through routes such as food and water and (3) personal variables such as cigarette smoking, past illness experience and genetic markers. Information about these covariates is used to compare like groups across exposure gradients and to compute response rates adjusted for differences which may influence the frequency of the response If area differences in response frequency are found, we can state with reasonable confidence that these differences are not due to other known epidemiologic determinants of the response.

TABLE 2

CHESS HEALTH INDICATORS

I. Acute Response

1. Daily symptom diary of asthmatics.

2. Daily symptom diary of subjects with chronic respiratory disease.

3. Daily symptom diary of subjects with cardiac disease.

4. Acute irritation symptoms (cough, eye irritation, chest pain) in families during "episodes."

5. Deviation from expected mortality (predicted from temperature, season and pneumonia - influenza deaths) during episodes.

II. Chronic Response

1. Pollutant burdens: concentrations of trace substances in tissue specimens, including hair, blood, urine, placentae, necropsy specimens.

2. Lung function (spirometry) of school children.

3. Bi-weekly acute respiratory disease incidence in families of elementary school children.

4. Frequency of lower respiratory disease episodes (croup, pneumonia, acute bronchitis) in children.

5. Prevalence of chronic respiratory disease symptoms in parents of school children.

ENVIRONMENTAL MONITORING

We establish an environmental monitoring station near the geographic center of study populations. In most cases, this setting places the residences of subjects within one and one-half miles of the monitoring station. In placing the station, we consider topography, land use adjacent to the study area and emission sources. Whenever possible, we place the inlet of the monitoring instruments at head level in an appropriate shelter removed from proximate sources of pollution such as road dust, if these sources are not common to the study area as a whole.

Table 3 lists pollutants, measurement methods and averaging times involved in the environmental monitoring portion of the CHESS program. To maintain both our acronym and the associated analogy, we call this phase CHESS-CHAMP (Community Health and Ambient Monitoring Program). We are currently collecting monthly samples and daily twenty-four hour samples for particulates and gases at 20 environmental monitoring stations. Continuous monitors are operating in some of the New York and Utah stations. To relate environmental variations to the acute response health indicators, we need the data from continuous monitors providing real time pollutant concentrations. A prototype continuous monitoring station having automatic acquisition with magnetic tape storage and "on-call" telemetry output of data is now under development and will be established in the entire CHESS-CHAMP system in the near future. We feel that the telemetry system is necessary on an "on-call" basis to enable routine checks on instrument performance, to process data on a daily schedule, and to provide access to current data during air pollution episodes.

Duplicate sampling and calibration of all instruments are to be obtained on a systematic basis and are critical to insure accuracy and consistency of instrument performance in the CHESS-CHAMP system.

STUDY STRATEGY

Selection of CHESS area sets and pollutant exposure is dictated by the existence of the six published air quality criteria. These include criteria for particulate matter, sulfur oxides, nitrogen oxides, photochemical oxidants, hydrocarbons and carbon monoxide. We do not include carbon monoxide in our pollutant exposure sets because we feel that short term exposure effects of carbon monoxide can be more precisely studied in controlled exposure chambers, while the effects of long-term area exposure to carbon monoxide are likely to be confounded with the effects of other emission products of vehicles. Furthermore, to date we have found no area set demonstrating a consistent exposure gradient for carbon monoxide.

TABLE 3

CHESS ENVIRONMENTAL MONITORING

I. Monthly samples (all stations)

1. Dustfall: trace substances, including: lead, cadmium, zinc, arsenic, mercury, and polychlorinated biphenyl compounds.

2. Composited daily high volume sampler filters: trace substances (as in #1).

3. Sulfation plates.

II. Twenty-four hour samples (all stations)

1. High volume sampler:
   -total suspended particulates.
   -suspended sulfates and nitrates.

2. Cyclone sampler: respirable particulates and trace substances.

3. Integrated gas bubbler: SO_2 (West Gaeke Method).

III. Continuous samples (under development for all stations)

1. Automatic data acquisition with magnetic tape storage:
   -SO_2 (coulometric method).
   -NO and NO_2 (coulometric method).
   -hydrocarbons (flame ionization).
   -ozone (chemiluminescent method).

IV. Meteorological samples (under development for all stations)

1. Hygrothermograph (temperature and relative humidity).
2. Wind speed and direction.

V. Other Environmental samples (under development)

1. Milk from wholesalers.
2. Tap water.
3. Rainfall samples.
4. Market basket samples.
5. Locally grown vegetables (where practical).
6. Rodents.
7. Vacuum cleaner house dust.
8. Additional dustfall sites in each residential area.
9. Soil samples.

We select from the community a stable, middle class residential sector having the desired pollutant exposure within the area set. We choose middle class neighborhoods because these are most prevalent and because the social class of families is more evenly distributed in such areas.

Through enrollment listings of elementary schools in selected neighborhoods, we recruit family units for participation in the acute upper, acute lower and chronic respiratory disease surveys and for episode panels. We contact these families through single-time questionnaires for prevalence surveys, weekly diaries for panel studies, and bi-weekly telephone calls for acute respiratory disease incidence studies. We measure lung function of children in the schools which, in spite of bussing, still represent a large proportion of neighborhood families with children of school age. We collect hair, urine and blood samples from selected households previously enrolled for other of our surveys. These households also provide a source of other environmental samples, including vacuum cleaner house dust, water and soil samples, and locally grown vegetables.

Listings of subjects for enrollment in the asthma, chronic respiratory and cardiac disease panels are obtained partially from results of the prevalence surveys and partially from patient listings of physicians. For mortality studies, we use data for the entire metropolitan area and compute expected mortality based on a model which takes into account season, daily temperature and reported pneumonia - influenza deaths. These models provide the local health agency with the capability to quantitate excess mortality during air pollution episodes as soon as data on current mortality counts become available.

As feasibility studies are successfully completed, we intend to monitor other health indicators. Among these we are considering tear lysozymes, red blood cell fragility and survival studies, placental enzymatic activity profiles, reversible lung function changes in subject panels, and exfoliative cytology of sputum samples.

In the CHESS program, we do not measure health characteristics in probability samples of the population. We have two reasons for this study strategy. Firstly, we are establishing air quality standards to protect segments of the population particularly vulnerable to the effects of environmental pollution. Secondly, our methods for enrolling subjects are far less costly and cumbersome than methods for obtaining probability samples. We focus our surveys on two segments of the population: susceptible groups and easily accessible groups. We try to match these groups within the same area set for characteristics which may influence the frequency of the health indicator; that is, we identify and measure the covariates of each health indicator. The ubiquity of middle class families with school children and of subjects with asthma, heart or lung disease allows us to be reasonably confident about the applicability of our findings to most communities in the United States.

The existence of pre-enrolled subjects in area sets provides us with a readily accessible group of concerned, cooperative middle class families with identified exposure to known pollutants. By monitoring health indicators over a broad spectrum of possible biological responses, and by simultaneously quantitating the environmental exposure, we feel that we are making a scientifically valid effort to achieve our stated objectives, namely, to evaluate existing standards, to quantitate pollutant burdens, and to document the health benefits of environmental control. The large cost of environmental control should justify the effort.

HUMAN POLLUTANT BURDENS

J.F. Finklea, D.I. Hammer, T.A. Hinners, C. Pinkerton

Community Research Branch, DER, BAPS, APCO, EPA

411 W. Chapel Hill St., Durham, N. C. 27701

INTRODUCTION

The "Toxic Substances Control Act of 1971" now being considered by Congress is the most recent Federal effort to assure that the health of our citizens and the quality of our environment will be protected from the adverse effects of residues from our industrial society (1). This legislation covers metallic pollutants and synthetic organic compounds not covered by prior Federal legislation dealing with air pollution, water pollution, pesticides, nuclear material, solid waste disposal, foods and food additives. Human pollutant burden patterns can become an important tool in our evaluation of environmental problems associated with toxic substances and pesticides.

By definition, a tissue carries a pollutant burden whenever it contains an environmental residue greater than that needed for optimum growth and development. Each of us has multiple pollutant burdens. Human pollutant burden patterns may serve as monitors of the environment, as indicators of biological response, as inputs into environmental standards development and appraisal, as channel markers for research, and as safeguards for recycling technology.

AS ENVIRONMENTAL MONITORS

Human tissues are dose integrators for environmental pollutants. Everyone maintains a series of involuntary personal dosimeters recording pollutant exposures for different time intervals. Those who would utilize these dosimeters must appreciate the underlying, interlinked exposure and metabolic factors that determine tissue burdens(2). For example, metal exposures usually involve a number of different chemical compounds impinging upon man through multiple environmental media. Our intake of cadmium involves air, water, food and tobacco smoke. Pollutant absorption is likewise a function of exposure route, physical form and chemical composition. Lead fume is rapidly and almost completely absorbed from the lung and lead bound to respirable particulates is largely absorbed. Conversely, lead bound to large suspended particulates fails to penetrate deeply into the respiratory tract and is shunted by the mucociliary apparatus to the gastrointestinal tract where absorption is relatively limited. In addition, organic lead compounds are much more readily absorbed and more toxic than inorganic lead compounds(3).

Distribution, storage and excretion kinetics are generally more complicated than those of exposure and absorption. Biological half lives of pollutants greatly differ; for mercury it is 80 days, for lead 10 years. Acute pollutant exposure and mobilization of residues from depots are usually measured by blood and excreta levels. Even then different metabolites may indicate different processes. Plasma DDD and urinary DDA are better indicators of recent exposure than plasma DDE which may be of dietary origin or recently mobilized from fat(4). Depot tissue may also record isolated acute exposures as shown by metaphyseal lead lines and hair arsenic bands(5,6).

Different tissues selectively concentrate different pollutant residues. Tissues with high lipid content like fat and brain serve as depots for Kr^{85}, alkyl mercury, chlorinated hydrocarbon pesticide and polychlorinated biphenyl plasticizer residues. Liver and kidney retain most of the body cadmium burden while lead, selenium and strontium concentrate in bones and teeth. Similarly, lung retains a number of pollutants including asbestos. When sampling living populations the epidemiologist must focus on more available specimens including hair, blood and excreta.

Pollutant burden appraisals have been applied to the metallic pollutants arsenic, cadmium, copper, lead, manganese, mercury, strontium and zinc(7-12). They should be equally useful for antimony, beryllium, chromium, nickel, selenium, silver, tin and vanadium. Human pollutant burden patterns have also been sketched for the persistent pesticides DDT, dieldrin, lindane and heptachlor(13). Pollutant burdens related to other synthetic organic compounds such as colorants, flavors, plastic products, rubber products, surfactants and organic intermediaries have received little attention except for the polychlorinated biphenyl compounds and, recently, the phthalates(14). Not all pollutants can be approached through pollutant burden studies. For example, organo-phosphate pesticide exposure is best assayed by measuring erythrocyte cholinesterase levels rather than a pesticide metabolite. Similarly, pollutant burden studies would not seem valid for oxidant air pollutants or sulfur dioxide. Carbon monoxide, nitrogen dioxide and suspended particulate air pollution occupy an intermediate position. Recent but not remote exposure of non-smokers to carbon monoxide and nitrogen dioxide might be assayed by carboxyhemoglobin and methemoglobin levels and particulates in lung have been quantitated at necropsy.

On occasion human tissues, serendipitously preserved, have furnished important insights into current pollution problems. Environmentalists have utilized museum specimens to elucidate pesticide, plasticizer and trace element pollution problems(15,16). There is a pressing need for human tissue banks to build an environmental flashback capability. There are several thousand metallic and synthetic compounds which potentially involve human exposures and human pollutant burdens. Infectious disease investigators encountered similar problems with unknown agents three decades ago and set up a network of sera banks which permitted them to assess rapidly the relationship of newly discovered microbes to older microbes, to disease syndromes and to the factors of age, sex, race, residence and time. A coordinated tissue bank program could overcome the obstacles inherent in standard tissue preservation. Freezing, lyophilization and irradiation seem the methods of choice with special care being devoted to containers. Less dramatic but most useful tissue banks would involve easily preserved tissues like hair, deciduous teeth and nail clippings collected from living populations whose environmental exposures are concomitantly measured.

AS INDICATORS OF BIOLOGICAL RESPONSE

Human pollutant burdens should be related to biological responses which may be placed in perspective by the several bench marks. In general, one thinks of five stages of increasing severity. First, a pollutant burden not associated with other changes. Second, a pollutant burden associated with physiologic changes of uncertain significance. Third, a pollutant burden associated with physiologic changes that are disease sentinels. Fourth, a pollutant burden associated with morbidity. Fifth, a pollutant burden associated with mortality. Even where pollutant burdens have been measured, only fragmentary information is available to link most burdens with biological responses. Our laboratory colleagues have rarely focused their attention upon biological response methods that can be applied to living populations. A concerted effort should be mounted to discern sensitive biochemical and physiologic indicators which may be then related to pollutant burdens.

The five stage biological response spectrum is a useful conceptual framework. There are, however, limitations other than information gaps. Not all disease processes fit comfortably into the response gradient, as illustrated by teratogenesis. Pollutant imprinting may also occur. Here the latency period for the response is longer than the biological life of the pollutant residue. In neurologic disorders associated with manganese poisoning, tissue manganese levels are not elevated. Similarly, exposure of neonatal animals to a variety of short-lived carcinogens may initiate late neoplastic changes.

Despite the many underlying complexities, useful deductions may be made from pollutant burden patterns. Prudence is mandatory when interpreting the general population patterns. A given pollutant burden might represent a minor risk for an industrial population of medically prescreened adults and a major risk for susceptible groups in the general population. Susceptible groups include the pregnant mother and her fetus, infants and children, the elderly, chronic disease cases and patients with genetic deficiencies of glucose-6-phosphate dehydrogenese and $alpha_1$ antitrypsin.

Age patterns can help identify special risk problems. PCB residue levels were shown to rise with age before dropping precipitously during the eighth decade of life(14). A superficial, unjustified interpretation might be that survivors have low levels and that the residue may be linked with a lethal disorder. In this case the oldest sample was quite small and thus their mean PCB residue level was not reliable. This study also illustrates the need to understand residue metabolism. PCB residues are most likely bound by plasma lipids and thus may be innocently associated with important atherosclerosis risk factors. The PCB data represent a cross-sectional view of the time factor. Another method, the cohort approach, involves repeated sampling of the same population over time. Cross-sectional studies actually include several age cohorts.

Residues may have different sex patterns as shown by DDT where residues in males exceeded those observed in females at every age. This male excess may have been due to greater exposure or differing metabolism. Ethnic group and residence may likewise effect pollutant burden patterns. Black children had total plasma DDT residues greatly exceeding those of white children. Rural black children actually equalled levels found in pesticide formulators who are the most heavily exposed occupational group. Clearly, these children are more likely to suffer any ill effects related to DDT residues(13). On the other hand, rural blacks less frequently exhibit plasma PCB residues than urban whites.

Differences between cities are also found, as illustrated by the study of environmental exposure and hair trace element levels(2). The non-essential trace metals arsenic, cadmium and lead followed an environmental exposure gradient while the essential trace metals copper and zinc, for which homeostatic mechanisms are well developed, did not significantly vary along a more limited exposure gradient. Pollutant burden surveys may also reveal individuals with residue patterns that warrant special investigation to abort clinical disease and to determine unusual environmental exposure routes.

AS INPUTS FOR STANDARDS DEVELOPMENT AND APPRAISAL

One method of environmental protection will be the promulgation and enforcement of ambient and emissions standards. Standards for pollutants characterized by direct exposure from a single environmental medium are set by considering control technology, the associations between pollutant levels and adverse effects and then allowing a safety factor. Sets of standards for pollutants impinging upon man through multiple environmental media are much more difficult to derive. One input into such standards will be the pollutant burden pattern of populations at greatest risk. For these pollutants, routine monitoring of population burdens may prove at least as important as environmental monitoring. Population burdens may also delineate a priority order for standards development.

Critical evaluation of each set of standards will be necessary and changes in human pollutant burdens will be one way to discern whether standards actually are achieving environmental quality goals. Increasing pollutant burdens would warn that environmental controls are inadequate. Should population burdens approach levels associated with clinical toxicity, environmental standards would have failed and emergency action would be indicated.

AS CHANNEL MARKERS FOR RESEARCH

Pollutant burden patterns can serve as channel markers for research. Ubiquitous pollutants, even those that appear benign in acute toxicity studies, might increase the risk or hasten the onset of chronic diseases. Demonstrable pollutant burdens in newborns and young children should alert investigators to teratogenic, carcinogenic and mutagenic hazards. Peak levels at puberty should signal concern for reproductive performance. Laboratory evaluation of these hazards will be difficult and expensive. Selection of chemicals for toxicologic study will be critically important. Pollutant burden studies should contribute to the selection process. Conversely, laboratory studies and environmental assays can help select which pollutant burdens should be studied and with what frequency.

AS SAFEGUARDS FOR RECYCLING

In the past, control technology has focused upon disposal but in the future much more emphasis will be placed upon recycling limited resources. The feasibility of recycling treated sewage water directly into municipal reservoirs and consuming food produced from solid wastes has been questioned by those who fear technology will not sufficiently safeguard the consumer. When recycling systems are utilized, appropriate human pollutant burdens should be monitored along with the recycled products.

AN OVERVIEW OF RESEARCH

The Bureau of Air Pollution Sciences in the Environmental Protection Agency currently mounts human pollutant burden projects of two types. One series of investigations is designed to determine what tissues are most suitable for monitoring. Here blood, urine, hair, excreta and placenta may be assayed for selected trace elements or PCB. Multiple tissue sets are also being collected at necropsy. Whenever possible, study groups are chosen because of their exposure to an appropriate pollutant gradient. The second series of investigations tests specific hypotheses.

Additional pollutant burden research might follow several paths. Attempts should be made to link pollutant burdens with refined laboratory indices of health impairment. One promising index is a placental enzyme activity profile which might be structured with two distinct components. One component should consist of enzymes induced or altered by specific pollutants while the other could include four or five enzymes which are rate limiting for selected metabolic cycles. Changes in the second component could serve as a sentinel for occult environmental problems. Other refined indices might include changes in serum metalloenzyme kinetics, minor aberrations in lymphocyte karyotypes, changes in reflex patterns and alteration of performance or behavior. A second path leads towards studies of pollutant interactions which may be antagonistic, additive or synergistic and a final path leads to predictive efforts designed to prevent the emergence of new pollutant burden problems.

SUMMARY

Human pollutant burden patterns should be a most useful research and monitoring tool in our quest for environmental quality.

REFERENCES

1. H.R. 5390. Toxic Substances Control Act of 1971, March 2, 1971

2. Hammer, D. I., Finklea, J.F., Hendricks, R.H., Shy, C.M., Horton, R.J.M. Hair Trace Metal Levels and Environmental Exposure. Amer. J. Epid. 93(2): 84-92, 1971.

3. Engel, R.E., Hammer, D.I., Horton, R.J.M., Lane, N.M., Plumlee, L.A. Environmental Lead and Public Health. EPA, APCO, AP No-90, March 1971.

4. Morgan, D.P., Roan, C.C. Chlorinated Hydrocarbon Pesticide Residue in Human Tissues. Arch. Environ. Health 20(4): 452-457, 1970.

5. Nelson, W.E., ed. Textbook of Pediatrics. W.B. Saunders Co., Philadelphia 1969, p. 1487.

6. Wintrobe, M.M., et al., ed. Harrison's Principles of Internal Medicine. McGraw Hill Book Co., New York, 1970, p. 1705.

7. Vallee, B.L., Ulmer, D.D., Wacker, W.E.C. Arsenic Toxicology and Biochemistry. Arch. Industr. Health 21:132-151, 1960.

8. Schroeder, H.A., Balassa, J.J. Abnormal Trace Metals in Man: Cadmium. J. Chron. Dis. 14:236-258, 1961.

9. Underwood, E.J. Trace Elements in Human and Animal Nutrition; Chapter 3, Copper; Chapter 6, Zinc; Chapter 7, Manganese. pp. 48-99, 157-186, 187-217, Academic Press, New York, 1962.

10. Schroeder, H.A., Tipton, I.H. The Human Body Burden of Lead. Arch. Environ. Health. 17:965-78, 1968.

11. Bidstrup, P.L. Toxicity of Mercury and Its Compounds. Elsevier Publishing Company, New York, 1964.

12. Gaffney, G.W., et al. Strontium-90 in Human Bone from Infancy to Adulthood, 1962-1963. Radiol. Health Data Reports 7:383-386, 1966.

13. Finklea, J.F., Priester, L.E., Sandifer, S.H., and Keil, J.E. South Carolina Community Pesticides Study Report. Unpublished, 1969.

14. Finklea, J.F., Creason, J. P. Priester, L.E., Hauser, T., and Hinners, T. Polychlorinated biphenyl residues in human plasma exposure - a major urban pollution problem? Abstract submitted for 99th annual APHA meeting, October 11-15, 1971.

15. Jensen, S. PCB as Contaminant of the Environment - History. PCB Conference, September 29, 1970, National Swedish Environment Protection Board, Research Secretariat, Solna, December 1970.

16. Jaworowski, Z. Stable Lead in Fossil Ice and Bones. Nature 217:152-153, 1968.

PESTICIDE EXPOSURE INDEX (PEI)

Julian E. Keil, M.S.,* John F. Finklea, M.D.,+
Samuel H. Sandifer, M.D.,* and M. Clinton Miller, Ph.D.*

*Medical University of South Carolina

+Environmental Protection Agency

INTRODUCTION AND BACKGROUND

Pesticide exposure indices may be of value and serve several purposes:

1. In selecting study groups.
   a. Prior to blood or other precise determinations.
   b. When precise analytical methods are not available.
2. Providing reference for data evaluation, e.g. exposure versus reaction.
3. Corroboration of laboratory determinations.
4. Establishing thresholds of permissibility (safety).

In studying the health effects of pesticides on humans, it is desirable to quantitate chemical exposure so that dosage estimates may be compared with clinical or biochemical response. Indeed, while chronic pesticide exposure is considered to be potentially toxic (4), proof of this has been slow in coming because of the lack of demonstrated cause and effect relationships in man.

Durham (4) reports that two general methods exist for measuring pesticide exposure: (a) A direct method involving the use of some mechanism to trap the toxic material as it comes in contact with the subject during his exposure period. (b) The indirect method involving some effect of the toxicant upon the exposed individual.

The direct method apparently is fraught with difficulties and the indirect method, while applicable with measurable substances and responses, cannot be used with many chemicals.

Studies by Selby et al (7) attempting to establish an index of pesticide exposure among the general population have not been successful because exposure was not discernible in the studied group. It is thought (2) that the principal source of exposure to the general population is from food (90%) and other environmental sources (10%); the amount from all sources being extremely small.

Personal interviewing was considered as a method to provide an index which would reduce to one number the potential pesticide exposure considering (a) the type material, (b) the concentration of material, (c) the occupation of the exposed individual and (d) the exposure time. To demonstrate its success and usefulness a PEI method should correlate well with a measurable parameter (such as plasma levels of a substance) and should be able to be used successfully as a predictor of a measurable parameter or response.

METHODS AND PROCEDURES

PEI Trial and Format

A study group from coastal South Carolina consisting of 78 occupationally exposed and 63 control volunteers was queried over a six month period about their DDT exposure. At each interview, a venous blood sample was drawn into a heparinized vacutainer for DDT and metabolite analysis.

Participants in the exposed group were rated by type and concentration of exposure as well as occupation (see Table 1). A numerical weighting was assigned each combination of chemical forms and occupations based on the theses that exposure somewhat followed these gradients.

Occupation: Formulator > PCO ~ Farmer > Supervisory personnel > Controls. (PCO = Pest Control Operator).

Chemical Form: Technical (100%) > EC WP > DC > Ready-to-use sprays ~ Dust ~ Baits > Granules > no exposure.

In establishing the categories or forms of materials it was considered that synthesizers or formulators, whether handling wettable powders (WP), dust concentrates (DC), or emulsifiable concentrates (EC), also must come in contact with the technical (usually approximating 80-100% in concentration) grade of material. Additionally the surfactants and/or solvent in WP, DC or EC (25% to 50% concentration) would assist in penetration of the active ingredient through the skin. Thus, the reason for the highest rating. Farmers or PCOs make sprays either from WP or EC and are

Table 1

OCCUPATIONAL AND MATERIAL CLASSIFICATION
FOR CALCULATION OF PEI

Technical, WP, DC, EC	Synthesizer or Formulator	4
Sprays (from WP or EC)	Farmers, PCO's	3
Dust, Baits	Formulator	3
Dusts, Baits	Farmers, PCO's	2
Granules	Formulator, Farmer	2
Supervisory Exposure		1

WP = Wettable Powder; DC = Dust Concentrate;

EC = Emulsifiable Concentrate; PCO = Pest Control Operator

exposed to materials intermediate in concentration between technical materials and the ready to use, usually dilute (0.25% to 1%) sprays. Therefore it would be logical to put them in a lower category. The classification follows the scheme of diminishing concentration and probable exposure from "4" to "1". The lowest exposure value is assigned to supervisory personnel such as managers and superintendents who have occasion to go into and through exposed areas, but whose actual exposure is minimal.

The preliminary factor (1-4) thus obtained was multiplied by the number of days an individual was exposed to a pesticide in the 30 day period preceding the interview. While exposure time might be 30 days, the maximum based on a 5 day week would not exceed 22 days.

The product of predominant exposure type (Table 1) and exposure time in days may then be divided by ten to obtain a smaller working number and an estimate of exposure potential. For practical and computer purposes so that only one digit would reflect the exposure estimate, Table 2 was constructed.

Table 2

PEI FROM PRODUCT OF EXPOSURE AND TIME

Exposure Range[1]	PEI
0.1 - 10.0	1
10.1 - 20.0	2
20.1 - 30.0	3
30.1 - 40.0	4
40.1 - 50.0	5
50.1 - 60.0	6
60.1 - 70.0	7
70.1 - 80.0	8
80.1 or above	9

[1]Product of Exposure Type and Time in days

ANALYSIS OF PLASMA

DDT was chosen as the confirmatory pesticide of choice because analytical procedures are well established and measurements quite precise. Analysis of plasma for pp'DDT and its principal metabolites pp'DDE and pp'DDD was by gas chromatography, following the Dale-Cueto procedure (3). The instrument used was a Micro-tek Model MT220 equipped with a Tritium electron capture detector, operated at 20-25 volts, D. C. The column packing was 27-OV-1/3% QF-1. The column was glass, six feet long and one fourth inch in diameter. Operating temperatures were: column 185 C; injection port 240 C, detector, 210 C. The carrier gas was high purity nitrogen, at a flow rate of 70 cc per minute. Confirmation by micro-coulometry and extraction p-values (1).

STATISTICAL COMPARISONS

The number of interviews and blood samples obtained from each individual varied but averaged about three per person. As a consequence a mean PEI and mean plasma chlorinated hydrocarbon value was calculated for each study participant. A linear correlation of the mean PEI and mean pp'DDT, pp'DDE, and pp'DDD was performed.

RESULTS

Pesticide Exposure Index correlated in a highly significant fashion ($p < .001$) with the actual occurrence and levels in plasma of pp'DDT and its metabolites, pp'DDE and pp'DDD. Among exposed and control participants DDT gave the highest correlation (r=.43), DDD was intermediate (r=.37), followed by DDE (r=.29). Table 3 gives the results of this linear correlation.

Table 4 presents a distribution of PEI by increasing exposure and demonstrates an accompanying increase in mean DDT and metabolite levels. The DDT, DDD, and DDE plasma levels for those reporting no exposure (PEI = 0) agree with other work (5) in general population.

DISCUSSION

Pesticide Exposure Index (PEI) is a useful and effective tool when studying groups occupationally exposed to pesticides.

While almost 70% (98) of the study group reported no DDT exposure, it provided a strong base with which to compare the group of 43 exposed to varying levels of DDT.

It is probably significant that pp'DDT correlates somewhat better than the other metabolites inasmuch as it is regarded as an indicator of recent exposure. Calculation of coefficients of determination (r^2) indicates that the lesser correlation of DDE is attributable to another source of variation. Individual rates and types of hepatic and bacterial degradation may vary. Indeed, work in progress indicates that there may be individualistic floral responses to DDT ingestion.

Table 3

PEI CORRELATION WITH PLASMA DDT AND METABOLITES

Pesticide	Number	Correlation Coefficient r	Significance
DDT	141	.4336	$p < .001$
DDE	141	.2934	$p < .001$
DDD	141	.3667	$p < .001$

Table 4

PEI AND PLASMA PESTICIDE LEVELS

Mean PEI	No.	Mean Plasma Pesticide Level		
		pp DDT ppb	pp DDE ppb	pp DDD ppb
0	98	4.7	6.6	.9
.1 - 2	30	7.2	9.1	.7
2.1 - 4	7	12.9	12.4	1.8
4.1 - 6	6	13.6	10.2	3.1

Also worth considering (see Table 4) is the effect that as PEI increases (from 2.1-4 to 4.1-6) the mean DDT and DDE levels tend to level off. This suggests the possibility of a DDT threshold level in plasma similar to that reported in fat by et al (6). Additional studies with increased numbers in the higher PEI range may provide further information along these lines.

In this study no PEI over 6 was determined. This may be explained by the seasonal nature of pesticide usage. While a number of individuals may have had a high PEI for a particular month (as high as 8), it was modified by a non-usage period.

Consideration should be given to safety practices employed, when estimating pesticide exposure. Surely a worker who regularly wears a respirator, gloves, uses other safety equipment and washes frequently will have less contact than an individual who utilizes no protective equipment and whose hygiene practices are substandard. Work is in progress on this facet and will serve to refine further the Pesticide Exposure Index.

PEI is relatively easy to calculate and is adaptable to field studies of groups occupationally exposed to pesticides. While this investigation has mainly tested PEI with DDT exposure, the authors postulate that it is equally applicable to other types of pesticides and chemical exposure.

SUMMARY

Pesticide Exposure Index (PEI) is a useful and effective tool when studying groups occupationally exposed to pesticides. PEI, calculated as the product of pesticide type, concentration, occupation of the exposed and exposure time, correlated in a highly significant fashion ($p<.001$) with the actual occurrence and levels in plasma of pp'DDT and its metabolites, pp'DDE and pp'DDD. Among 78 occupationally and 63 control participants DDT gave the highest correlation ($r=.43$), DDD was intermediate ($r=.37$), followed by DDE ($r=.29$). Calculation of coefficients of determination indicates that the lesser correlation of DDE is attributable to another source of variation. This study was conducted over a period of one year and included three observations per person. Analysis of plasma was by gas chromatography.

REFERENCES

1. Bowman, M. C., and Beroza, M. "Extraction p-values". Analytical Chemistry, 37:291-292, 1965.

2. Campbell, J. E., Richardson, L. A., and Schaefer, M. A. "Insecticide Residues in the Human Diet". Arch. Environmental Health, 10:831-836, 1965.

3. Dale, W. E., Curley, A., and Cueto, C. "Hexane Extractible Chlorinated Insecticides in Human Blood". Life Science, 5:47, 1966.

4. Durham, W. F. "Assessment of Environment Exposure to Pesticides". Proceedings from the short course on occupational health aspects of pesticides. Edited by Ling and Whitaker, University of Oklahoma Press, Norman, Oklahoma, 1964.

5. Finklea, J. F., Keil, J. E., and Priester, L. E. Medical University of South Carolina. Unpublished data, 1969.

6. Hayes, W. J., Jr., Quinby, G. E., Walker, K. C., Elliott, J. W., and Upholt, W. M. "Storage of DDT in People with Different Degrees of Exposure to DDT". American Medical Association Arch. Ind. Health, 18(5):398-406, 1958.

7. Selby, L. A., Newell, J. W., Waggenspack, C., Hauser G. A., and Junker, G. "Estimating Pesticide Exposure in Man as Related to Measurable Intake; Environmental Versus Chemical Index". American Journal Epidemiology, 3:241-253, 1969.

SUMMARY

A Combined Index for Measurement of Total Air Pollution: Effects of Changing Air Quality Standards

Lyndon R. Babcock, Jr.

Department of Energy Engineering
University of Illinois at Chicago Circle
Chicago, Illinois 60680

ABSTRACT

Combined pollution indexes are useful for assessing overall air pollution levels and seem to be an overdue necessity. Formulators of pollution control policy must be able to trade off one pollutant against another, always being aware of the effect of a change in one pollutant level on overall air quality. An index called "pindex" is proposed in which six pollutants, a synergism, and photochemical effects are combined to yield a measure of total air pollution. Pindex combines the more generally accepted tenets of air pollution technology and, despite definite limitations, could be a useful tool for assessing total air pollution levels.

Within the pindex calculation, there are nine somewhat arbitrary weighting coefficients. Six coefficients are based on ambient air quality standards proposed for California,[5] while the remainder involve photochemistry and sulfur oxide-particulate synergism. This present paper extends earlier work[1] by specifically examining these coefficients. Implications of changes in certain air quality standards are discussed as they affect pindex results.

INTRODUCTION

Evaluating overall air pollution can be a complex undertaking. Urban air pollution consists of an often ill-defined mixture of several pollutants emitted from different energy and industrial processes. Additional secondary pollutants are created in the atmosphere. Synergisms can occur between certain pollutants. Despite these complexities, efforts must be made to total the effects of the individual pollutants. This paper presents such an attempt. Only with a usable total air pollution yardstick can we get the most from pollution control expenditures. In addition, a total air pollution measure enables one to compare overall air pollution in different localities.

Overall air pollution measures serve at least two purposes. First, they can be used to give the layman a more meaningful assessment of air pollution severity. The layman wants to know "how bad it is." He may even object to the control agency hiding behind individual part-per-million numbers which the layman does not understand. Several partial indexes are now in use by specific control agencies.

Second and perhaps more important, a combined air pollution measure or index enables evaluation of the tradeoffs involved in alternative air pollution control policies or in evaluation of control equipment which, for instance, reduces levels of certain pollutants while increasing levels of others. The combined air pollution index, called pindex, can serve as both a lay information tool and as a technical tool.

In an earlier paper[1] pindex was described and applied to several examples. The method and two of the examples are reviewed herein before showing the effects of altering some of the pindex coefficients.*

The simplest measure of total air pollution unfortunately is in wide use and involves the simple summation of individual pollutant emission weights. A familiar summary of USA emissions[2] is shown in the upper half of Table 1. Quite clearly, when using an uncorrected weight basis, the USA air pollution problem consists largely of carbon monoxide emanating from automobile engines. Note that oxidant does not appear in Table 1. Worse, relative toxicities are not considered in gross weight comparisons.

---

*Several figures and portions of text included herein originally appeared in the writer's earlier paper[1] and are reproduced by permission of the Editor, Journal of the Air Pollution Control Association.

In addition to material reviewed here, the earlier paper compared pindex to other indexes, included a sample calculation to obtain a pindex value, and applied the pindex method for comparison of ambient air pollution levels in 10 USA cities.

Table I. USA air pollution source distribution[2]

	PM	SO_x	NO_x	CO	HC	Total
Uncorrected Basis (10^6 tons/year)						
Transportation	1.8	0.5	3.1	59.6	9.7	74.7
Industry	6.0	8.7	1.6	1.8	3.7	21.8
Power plants	2.4	10.2	2.4	0.5	0.1	15.6
Space heating	1.2	3.4	0.8	1.8	0.5	7.7
Refuse incineration	0.6	0.2	0.1	1.3	1.0	3.2
Total	12.0	23.0	8.0	65.0	15.0	123.0
Pindex Levels (percent of grand total)						
Transportation	6	1	7	2	3	19
Industry	22	11	4	0	1	38
Power plants	11	12	6	0	0	29
Space heating	5	4	2	0	1	12
Refuse incineration	2	0	0	0	0	2
Total	46	28	19	2	5	100

Another example deals with jet engines, a relatively recent source of air pollution. George, Verssen and Chass[3] compared the effects of fuel type and burner can design upon emissions from 4-engine jet aircraft. Some of their results are shown in the upper half of Table II. These authors summarized their work in part as follows:

> Substitution of the new "smokeless" burner cans . . . reduced emissions of total air contaminants from this engine by nearly 75 percent. Emissions of particulates were reduced by 23 percent, carbon monoxide by about 23 percent, and hydrocarbons and organic gases by 99 percent; nitrogen oxides showed an increase of about 40 percent.

Table II. Emissions from 4-engined jet aircraft[3]

	PM	SO_x	NO_x	CO	HC	Total
Uncorrected Basis (pounds/flight)						
Uncontrolled	19.3	4.0	12.4	26.3	172.8	234.8
With JP-4 fuel	12.3	2.8	12.9	31.9	37.0	96.9
With "Clean" burner cans	14.9	4.0	17.4	20.4	1.1	57.8
Pindex Level (normalized)						
Uncontrolled	53	4	26	1	16	100
With JP-4 fuel	34	3	27	1	9	74
With "Clean" burner cans	42	4	35	0	2	83

Table III. Derivation of tolerance factors

	Proposed California Standards [5]	Tolerance factors (ppm)	Tolerance factors ($\mu g/m^3$)
Oxidant	0.1 ppm for 1 hr	0.10	214
Particulate matter	Visibility below 7.5 miles for 12 hr, below 3 miles for 1 hr	—	375
Nitrogen oxides	0.25 ppm for 1 hr (for nitrogen dioxide)	0.25	514
Sulfur oxides	0.1 ppm for 24 hr 0.5 ppm for 1 hr (for sulfur dioxide)	0.50	1430
Hydrocarbons	—	—	19300
Carbon monoxide	20 ppm for 8 hr	32.0	40000

First, what is the most serious emission from a jet engine? Second, how much of the commendable reduction in hydrocarbons is counteracted by the increased emission of nitrogen oxides? Questions such as these should be answered before wholesale engine conversions take place. Similar tradeoffs exist for some automobile emission control devices.

AIR QUALITY STANDARDS

Air pollution standards can provide the basis for combined pollution indexes. There is fair agreement as to the levels at which individual pollutants begin to endanger the breathing public. The NAPCA air quality criteria[4] attempt to summarize the effects of individual pollutants. Many agencies have now adopted air quality standards. Standards proposed for the State of California[5] in 1969 are representative and are shown on Table III. These standards are fairly comprehensive, but note that no standard was proposed for hydrocarbons, and there is no common time basis.

Presumably, when the ambient level of a given pollutant exceeds the standard, a dangerous or unhealthy situation exists, and some corrective action is taken. Standards then give us a basis for comparison of different pollutant concentrations. In California, for example, equivalent unpleasantness or severity was assumed to occur at 0.5 ppm sulfur dioxide or 0.25 ppm nitrogen dioxide or 0.1 ppm oxidant. Unfortunately such standards often say nothing about combinations of pollutants. Even aside from the synergism question, could 0.24 ppm nitrogen dioxide <u>and</u> 0.09 ppm oxidant together (both below the standard) be worse than 0.51 ppm sulfur dioxide alone (above the standard)?

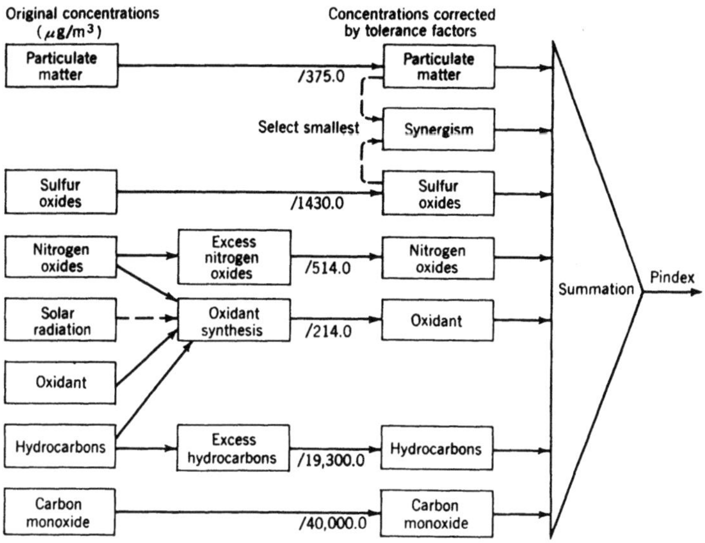

Figure 1. Pindex calculation scheme

EXISTING COMBINED INDEXES

There have been relatively few attempts to go beyond the standards step. Yet it seems logical to use the standards to arrive at a combined pollution level. The few existing combined indexes are specialized, being based on specific existing instrumentation or upon specific local situations. Most existing indexes include only a few pollutants.

PINDEX, A COMBINED AIR POLLUTION INDEX

Pindex was created to serve as a more all-encompassing index. It sums the contributions of all the major combustion emissions: particulate matter, sulfur oxides, nitrogen oxides, carbon monoxide, and hydrocarbons. In addition, provisions for oxidant as either a primary or secondary pollutant and a term representing particulate matter-sulfur oxides synergism are included.

In order to use standards as weighting factors, it is necessary to adjust all standards to the same units and time basis. The assumptions and methods used to arrive at the pindex factors were described earlier.[1] These factors are listed in the last column of Table III and show an enormous variation. For example, based on the proposed California standards, pindex assumes that 214 μg/m^3 of oxidant is as equivalently unpleasant as 40,000 μg/m^3 of carbon monoxide. A factor of 200 separates the two pollutants.

The calculation scheme is shown schematically on Figure 1. The inputs include the raw emissions plus solar radiation which influences the amount of hydrocarbons and nitrogen oxides converted to oxidant. Alternatively, oxidant can be considered an input as in ambient air quality data. Next the revised concentrations of pollutants are weighted by their tolerance factors. Finally each corrected pollutant plus the particulate-sulfur oxides synergism term are summed to yield the pindex level.

RESULTS

Pindex levels were determined for the examples mentioned earlier:

USA Source Distribution

USA nationwide emissions, before and after adjustment by pindex, are summarized in Table I. The pindex results were normalized such that the grand total is 100 percent. Although not shown separately, the oxidant and synergism terms were calculated and added back into their respective precursors.

Use of tolerance factors plus inclusion of oxidant and synergism terms have completely reordered the USA air pollution problem. Carbon monoxide, which dominated the source distribution based on emission weights, became almost insignificant after pindexing. Particulate matter became the most serious USA air pollution problem, with sulfur oxides second, and nitrogen oxides a strong third. Despite its essential contribution to photochemical oxidant synthesis, hydrocarbons have assumed a low value only slightly ahead of carbon monoxide.

Among the sources, transportation remains a significant pollution category, but pindex lowered its ranking to third behind both industry and power plants.

Sawyer and Caretto,[6] working independently of the writer, came to the same conclusion at almost the same time. Using somewhat different weighting factors, they distributed source importance for the same emission data as follows:

	percent
Motor Vehicles	12
Industry	37
Power Plants	36
Space Heating	10
Refuse Disposal	5

Table IV. USA air pollution source distributions (as weighted by pindex)

	1964	1968
Transportation	19	16
Industry	38	18
Power plants	29	--
Space heating	12	--
Refuse incineration	2	3
Stationary combustion	--	45
Miscellaneous	--	6
Forest fires	--	12
	100	100

The pindex and Sawyer and Caretto comparisons are based on emission data from the early sixties. A 1968 NAPCA survey[7] has recategorized some of the emissions and added forest fires as a major source. Total emissions have increased 75 percent from 123 to 214 million tons/yr. Hopefully most of the increase can be attributed to more sophisticated survey methods rather than to increased emissions.

The old and new, after pindex correction, are compared on Table IV. Note that the industrial contribution is reduced in the 1968 survey because stationary combustion emissions have been removed from the industrial category and combined with emissions from space heating and electric power plants. In any event, pindex correction of the 1968 data has further reduced the importance of the transportation category. Indeed, if trucking, aircraft and other non-automobile transportation emissions were removed from the pollution contribution of the remaining private automobile sector would be reduced to below 10 percent. Such findings might bring joy to the automobile industry, whereas personnel in other

Table V. EPA proposed air quality standards, $\mu g/m^3$

	Annual	24hr	8hr	3hr	1hr	Pindex (1 hr)
Oxidant	--	--	--	--	125	214
Particulates	75 (60)	260 (150)	--	--	--	375
Nitrogen oxides	100 (100)	250 (250)	--	--	--	514
Sulfur oxides	80 (60)	365 (260)	--	--	--	1430
Hydrocarbons	--	--	--	125	--	19300
Carbon monoxide	10000	--	15000	--	--	40000

industries and power plants might begin to question the tolerance factors.

The tolerance factors may well be incorrect. Indeed, the standards in California have since changed, and now the federal Environmental Protection Agency (EPA) has proposed nationwide ambient air quality standards for adoption during 1971.[8] The proposed EPA standards are compared with those used in pindex in Table V.

Primary (health related) as well as secondary standards have been proposed, but after adjustment to a common time basis, agreement with the pindex ranking would be fairly good with one exception. EPA specifies the more important "reactive" rather than the very much larger, but less meaningful, "total" hydrocarbon standard used in pindex. Unfortunately very little reactive hydrocarbon data is available either for sources or for ambient air in cities. It seemed unwise to revise the pindex hydrocarbon factor until reactive hydrocarbon analyses become more common.

Air quality standards can be expected to change frequently in the future as more is learned about pollutant toxicity and other undesirable pollutant effects. What then is the expected effect of such changes upon pindex results? First, note that a concurrent tightening of all the standards need have no effect, unless one standard were significantly changed relative to the others.

Figures 2 - 8 illustrate the effects on the 1964 USA pollution distribution as single standards or tolerance factors are changed. The results are normalized such that the total equals 100 percent for all values of the tolerance factor being studied. The verti-

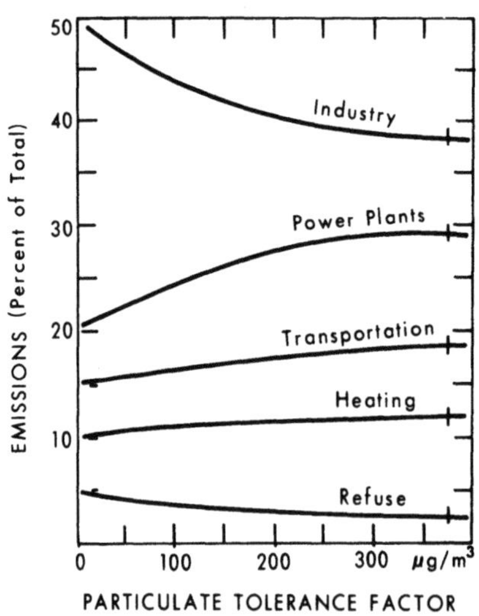

Figure 2. Particulate tolerance factor and USA sources

cal marks on each curve indicate the standard pindex tolerance level (i.e., where transportation equals 19 percent, industry equals 38 percent, etc.; see lower half of Table I). Finally, notice that the pollutant in question becomes less unpleasant and less significant as the tolerance factors increase toward the right sides of Figures 2 - 8.

Particulates (Figure 2). Particulates are the dominant industrial emission; thus, the industrial contribution increases rapidly as particulate tolerance is reduced. Also in Figure 2, the power

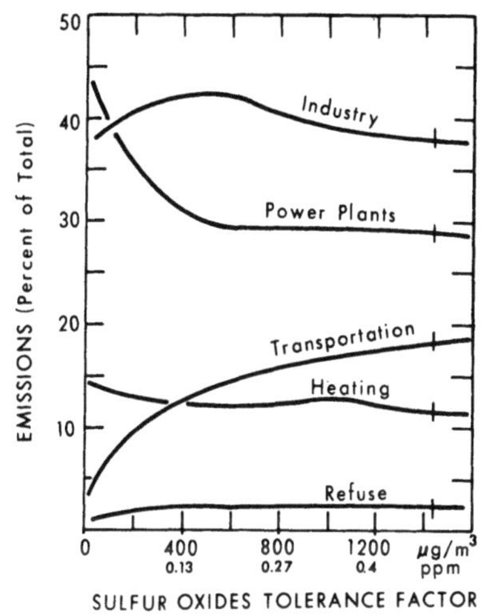

Figure 3. Sulfur oxides tolerance factor and USA sources

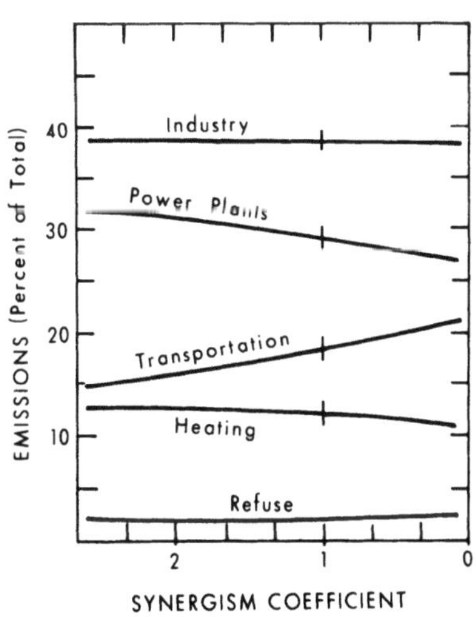

Figure 4. Synergism coefficient and USA sources

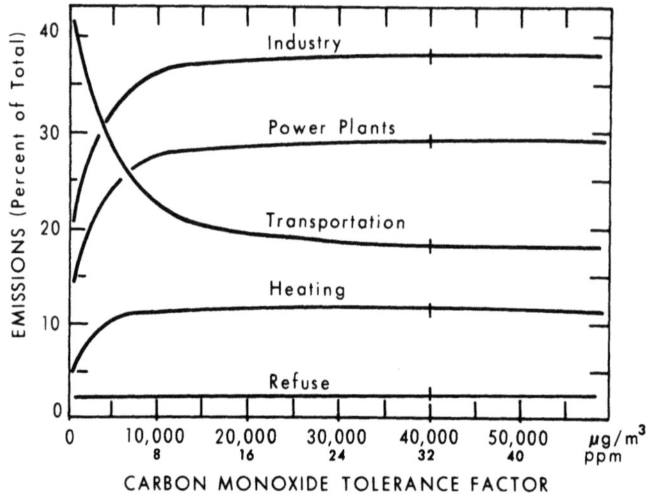

Figure 5. Carbon monoxide tolerance factor and USA sources

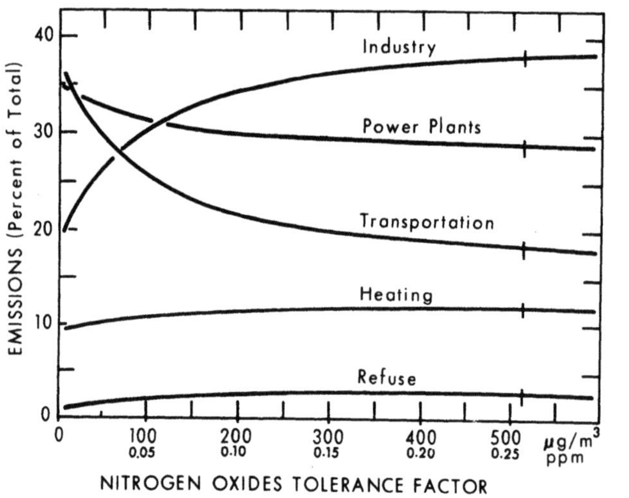

Figure 6. Nitrogen oxides tolerance factor and USA sources

plant contribution reaches a maximum, as luck would have it, near the standard tolerance factor level. Finally, the ranking is not altered (the curves do not intersect) even as the particulate tolerance approaches zero.

Sulfur oxides (Figure 3). Power plants are the major contributors of sulfur oxides; reduced sulfur oxides tolerance increases the relative pollution contribution of power plants. The transportation category emits little sulfur oxides, and thus transporta-

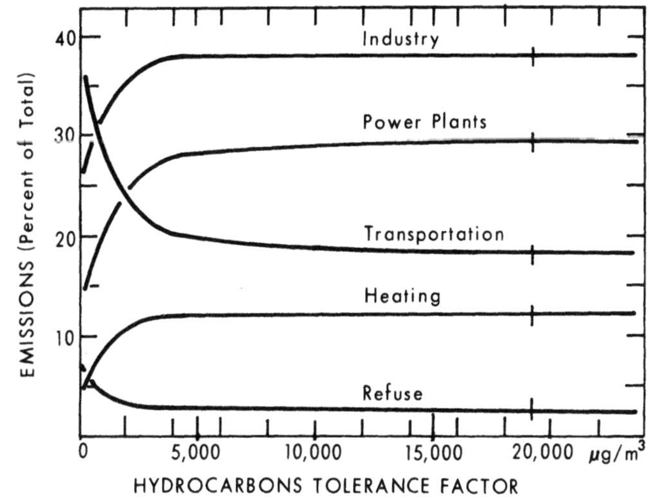

Figure 7. Hydrocarbons tolerance factor and USA sources

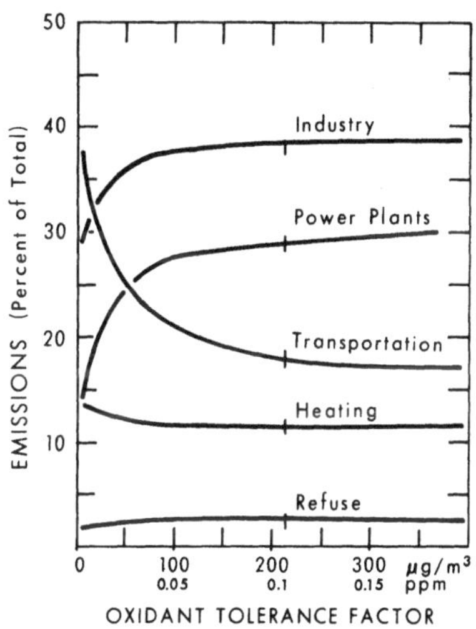

Figure 8. Oxidant tolerance factor and USA sources

tion increases in relative importance as sulfur oxides are de-emphasized (tolerance factor increased).

The initial rise in the industrial curve is a response to the very rapid decrease in power plant contribution. But industry is also a significant source of sulfur oxides and, after the maximum is reached, the industrial sector also benefits from further reductions in the tolerance factor. Note the smaller maximum for space heating and also note the large decrease in tolerance factor required to rearrange the categories.

Sulfur oxides-particulate synergism (Figure 4). The synergism coefficient is one of the more questionable in pindex; as the relationship is altered, power plants are most affected, having nearly equal amounts of each pollutant. Transportation effects are relatively immune to changes in the synergism term; the transportation contribution rises to compensate for the decreasing power plant category.

Carbon monoxide (Figure 5). The USA source distribution is extremely insensitive to changes in carbon monoxide tolerance factor. Reduction of the factor to near zero would be required to make transportation the dominant category.

Nitrogen oxides (Figure 6). Similarly, the source distribution is insensitive to changes in the nitrogen oxides tolerance factor.

Hydrocarbons (Figure 7). The source distribution is also very insensitive to changes in the hydrocarbon tolerance factor. Note that a "reactive" hydrocarbon emission survey will be needed in order to evaluate the proposed EPA air quality standard for reactive hydrocarbons.

Oxidant (Figure 8). Pindex factors such as solar coefficient and solar radiation yield relationships similar to that for oxidant. Reduced oxidant tolerance and/or increased solar radiation (which increases oxidant level) increase the importance of the transportation category, with this category dominating when oxidant only is considered. The photochemical assumptions in pindex warrant further study. Hopefully improved relationships can be inserted without unduly complicating the model.

Emissions from Jet Aircraft

Emission levels of jet aircraft before and after the pindex corrections are compared in Table II. On a weight basis, hydrocarbons accounted for 74 percent of the uncontrolled total. Use of JP-4 fuel significantly reduced hydrocarbon emissions. The use of "clean" burner cans effectively eliminated hydrocarbon emissions but increased nitrogen oxide emissions. Unfortunately, the pindex evaluation indicates that the particulates and nitrogen oxides are the most important jet engine emissions with hydrocarbons amounting to only 16 percent of the total.

In the pindex comparison, use of JP-4 fuel caused significant reductions in both the particulate level and the hydrocarbon level without increasing the nitrogen oxides problem. The "clean" burner

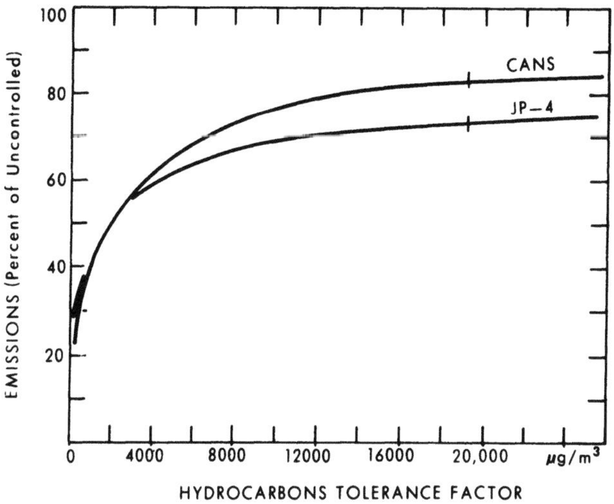

Figure 9. Hydrocarbons tolerance factor and jet engine emissions

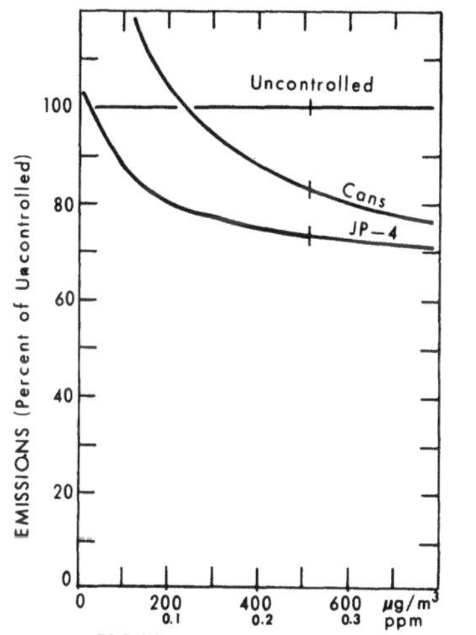

Figure 10. Nitrogen oxides tolerance factor and jet engine emissions

cans were less effective in reducing particulates, and 64 percent of the commendable hydrocarbon reduction was neutralized by the increase in nitrogen oxides. Pindex, at least, suggests preference for the JP-4 fuel. It would be of interest to obtain emission data for the combination wherein JP-4 fuel was used in an engine equipped with "clean" burner cans.

Changes in tolerance factors also alter the relative desirability of burner cans versus JP-4 fuel. The most interesting

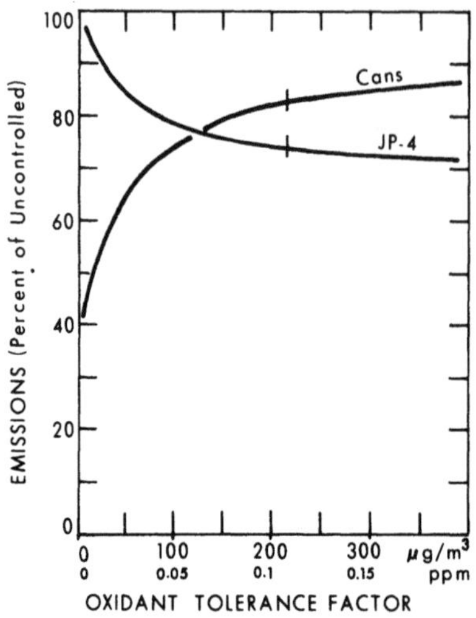

Figure 11. Oxidant tolerance factor and jet engine emissions

relationships are summarized on the figures 9 - 11. For these comparisons, the pindex revised emissions are normalized such that the emissions from uncontrolled engines always equals 100 percent, whatever the tolerance factor.

Hydrocarbons (Figure 9). The "clean" burner cans only reduce engine pollution by 15 percent when the standard tolerance factors are used. Reducing the hydrocarbon factor, of course, remarkably increases the desirability of the burner cans. Even so, the JP-4 and burner-can curves cross at a very low hydrocarbon tolerance factor.

Nitrogen oxides (Figure 10). The nitrogen oxides factor is also significant here; if the reduction in nitrogen oxides tolerance were large enough, the pollution from "clean" burner cans would surpass that from uncontrolled engines.

Oxidant (Figure 11). The oxidant relationship is interesting and is presented here even though it is unlikely that jet aircraft would represent the exclusive supply of both oxidant precursors (hydrocarbons and nitrogen oxides). Figure 11 indicates that reduced oxidant tolerance level decreases the desirability of JP-4 fuel. The slight increase in nitrogen oxides coupled with insufficient reduction in hydrocarbons can yield oxidant concentrations close to that emitted from uncontrolled engines. For the "clean" burner cans, however, insufficient hydrocarbons remain to react with the increased nitrogen oxide emissions. Oxidant formation is thus reduced, and the burner cans become more desirable as oxidant tolerance factor is reduced.

Figure 11 indicates that, where synergisms or secondary reactions occur after emission, the importance of a single source may well depend on synergistic emissions from other sources within the same air shed.

IMPLICATIONS

The sensitivity analysis reported here helped define many areas within our air resource management system which merit further study. Changes in the tolerance factors can indeed affect the pindex results. Yet, in general, the work supports the validity of pindex. Rather large changes in coefficients can be made without significantly altering the pindex-derived conclusions reported herein.

Thus pindex enables one to rationally determine the contributions of individual pollutants to the overall air pollution problem. If the results are meaningful, such a tool can assist in determining air pollution control priorities. As in the examples described, priorities may be somewhat off target already. Programs to reduce carbon monoxide emissions from automobiles or hydrocarbon emissions from jet aircraft may be disappointing, even if the control goals are attained. Pindex doesn't say such pollutants should be neglected, but it points to more important pollutants which may deserve higher priorities.

Pindex evaluations showed particulate matter to be our most serious air pollution problem. On Table I, particulate matter accounted for 46 percent of pindexed USA emissions. Even in the transportation category, particulates amounted to one-third of the total. On Table II, particulate matter accounted for 53 percent of the jet engine pindexed total. Despite long efforts and well understood technology, particulate matter remains our most serious air pollution problem.

Concern about nitrogen oxides surfaced more recently. Current attention given to nitrogen oxides seems well warranted. Nitrogen oxides (especially nitrogen dioxide) are toxic, and according to pindex, nitrogen oxides are often the limiting reactants in oxidant synthesis. If true, it may be necessary to obtain nearly complete removal of hydrocarbons before oxidant levels will decline. Alternatively, any decrease in nitrogen oxides should show a proportionate decline in oxidant levels.

CONCLUSIONS

Combined pollution indexes seem to be an overdue necessity. Formulators of pollution control policy must be able to trade off one pollutant against another, always being aware of the effect of a change in one pollutant level on overall air quality.

Pindex combines the more generally accepted tenets of air pollution technology and, despite definite limitations, could be a useful tool for assessing total air pollution levels. The merit of pindex will be measured only when it is put to use. Some may consider the method to be grossly oversimplified, but complexity should be added only when shown to be necessary. Hopefully pindex is a versatile framework, general enough so that it may be updated as its shortcomings become apparent and as new information becomes available. In addition to revising existing factors and relationships, provision for additional pollutants might be added in the future. The recent work reported here indicates that pindex results are surprisingly insensitive to changes in the tolerance factors. Future studies will investigate other facets of the model.

ACKNOWLEDGMENTS

The initial work was carried out at the University of Washington, Seattle, with financial support provided by the National Air Pollution Control Administration U. S. Public Health Service through Air Pollution Special Fellowship No. F3 AP 41,084. Recent financial support has been provided by the University of Illinois at Chicago Circle and by National Science Foundation Grant No. GK 27772.

REFERENCES

1. Babcock, L. R. "A Combined Pollution Index for Measurement of Total Air Pollution," J. Air Poll. Control Assoc., 20:10, 653-659 (October 1970).

2. National Academy of Sciences - National Research Council. Waste Management and Control (Publication 1400). Washington (1966).

3. George, R. E. et al. "Jet Aircraft: A Growing Pollution Source," J. Air Poll. Control Assoc., 19:11, 847-855 (November 1969).

4. National Air Pollution Control Administration (NAPCA). Air Quality Criteria for . . . (AP-49, 50, 62, 63, 64). Durham, North Carolina (1969-1970).

5. California State Department of Health. Recommended Ambient Air Quality Standards (May 21, 1969).

6. Sawyer, R. F. and Caretto, L. S. "Air Pollution Sources Reevaluated," Environ. Sci. Technol., 4:6, 453-455 (June 1970).

7. Secretary of Health, Education and Welfare. Progress in the Prevention and Control of Air Pollution (Third Report to Congress: 91st Congress, 2nd session, Doc. No. 91-64). Washington (1970).

8. "Environmental Protection Agency Proposes National Air Quality Standards," J. Air Poll. Control Assoc., 21:3, 149 (March 1971).

RATIO OF SULFUR DIOXIDE TO TOTAL GASEOUS SULFUR COMPOUNDS AND OZONE TO TOTAL OXIDANTS IN THE LOS ANGELES ATMOSPHERE - AN INSTRUMENT EVALUATION STUDY

Robert K. Stevens,* J. A. Hodgeson,* Lewis F. Ballard,+ and Clifford E. Decker+

*U. S. Environmental Protection Agency, Technical Center, Research Triangle Park, N. C. 27711
+Research Triangle Institute, Research Triangle Park, N. C. 27709

ABSTRACT

During the past 3 years, several promising air pollution monitoring systems have been developed by APCO and tested on a limited scale. These systems, however, have not been tested in enough geographical areas to determine the effect of various combinations of pollutants in urban environments on instrumental response. This report describes a comprehensive evaluation program designed to compare these newly developed instruments with classic procedures for measuring ambient concentrations of sulfur dioxide and ozone. The following sulfur dioxide monitoring principles were compared: (1) colorimetric (West-Gaeke), (2) conductometric, (3) coulometric, (4) flame photometric, and (5) combined gas chromatographic-flame photometric. A colorimetric analyzer and a coulometric system were used as oxidant monitors; while a modified Regener and Nederbragt system were used to measure ozone. The evaluation was performed in Los Angeles, California over an 83-day period between September and December 1970. The major results of this indicate that: (1) the flame photometric detector (FPD) and gas chromatographic-FPD instruments required the least maintenance and were operational for more than 93 percent of the evaluation period, (2) coulometric and conductometric SO_2 measurements were consistently higher than values obtained by flame photometric measurements,

(3) hydrogen sulfide and methyl mercaptan were occasionally present in the Los Angeles atmosphere, (4) sulfur dioxide frequently exceeded 0.030 ppm during the afternoon; and, (5) total gaseous sulfur in Los Angeles was better than 90 percent sulfur dioxide. Ozone and oxidant measurements corrected for NO_2, are equivalent at peak concentrations during the daylight hours in downtown Los Angeles.

INTRODUCTION

Rationale -- The Air Pollution Control Office over the past 4 years has developed several new analytical methods for measuring ambient concentrations of sulfur dioxide, hydrogen sulfide, and ozone. In addition, commercial manufacturers have also developed improved sensors for measuring these pollutants. To evaluate the performance of these new instrumental techniques, a comprehensive field study embracing different geographical areas must be conducted in order to determine the effects on the response of the analyzers of various combinations of pollutants in an urban environment.

The ultimate aim of this continuing program is to establish on an absolute and a comparative basis the degree to which the instruments evaluated meet the needs of air pollution control agencies. This study is divided into two phases. One phase deals with the evaluation of the response characteristics of sulfur dioxide and ozone monitors and the second phase objectives deal with determining the relationship of sulfur dioxide to total gaseous sulfur and ozone to total oxidants.

Previous Studies -- In 1968 Stevens, O'Keeffe, and Ortman,[1] suggested that the total gaseous sulfur in most urban atmospheres is in the form of sulfur dioxide. They based their hypothesis on the knowledge that millions of tons of sulfur dioxide are emitted annually from power plants and a variety of industrial activities, whereas the total emissions of other sulfur gases is only a small fraction of the SO_2 inventory. A few short-term measurements were made in Cincinnati, Ohio, and Raleigh, N. C., utilizing a gas chromatograph[2] designed to produce a specific response to SO_2. These studies indicate that in these cities, measuring total gaseous sulfur in the atmosphere was equivalent to measuring SO_2.

Palmer, Rodes, and Nelson[3] evaluated the performance of commercially available sulfur dioxide monitors over a 3-month period in New York City. Their study was designed to determine the performance of these instruments under conditions closely related to field operation. The Palmer study[3] measured such characteristics as calibration drift, maintenance requirements, and variation in response for each air monitor. The mean concentration of sulfur dioxide detected by all instruments included in the study over the 3-month period was 0.206 ppm.

Virtually no comparative studies between chemiluminescent ozone monitors and oxidant analyzers have been performed. This is due to the relatively recent development and availability of specific ozone monitors.

Site Selection -- The Palmer study[3] was performed during the winter when sulfur dioxide concentrations are relatively high and the concentration of photochemically generated species, such as ozone, should be low.

Analysis of the data presented in the Palmer study revealed that when the mean concentration of SO_2 in the atmosphere was above 0.164 ppm, most of the monitor responses agreed reasonably well. The report did not indicate whether the agreement would hold at mean concentrations lower than those observed in New York City nor did it indicate what other atmospheric factors might affect the performance of the analyzers.

To answer these questions, and to compare some of the new developments in SO_2 monitoring instrumentation, Los Angeles, California, was selected as the initial location at which to perform the evaluation.

A mobile laboratory trailer was considered the most realistic vehicle for transporting and housing the instruments for the evaluation. Figure 1, a diagram of the mobile laboratory, shows the location of the monitoring instruments, data acquisition system, and gas sampling manifold. The trailer was located adjacent to the Los Angeles County Air Pollution Board's central research facility on San Pedro Street in downtown Los Angeles.

Instrument Selection -- Classical colorimetric (West-Gaeke), coulometric, and conductometric analytical techniques were used as reference methods for evaluating the flame-photometric-detector (FPD), total-sulfur monitor and the gas chromatographic-FPD (GC-FPD) system. The commercial instruments selected by the authors as representative of the analytical principles to be evaluated are included in Table I.

All of the instruments are commercially available and, with the exception of the Tracor GC-FPD analyzer, are being used by a variety of state and Federal agencies. The automated GC-FPD combination is a relatively new technique and, therefore, is being used by only a small number of state and Federal agencies.

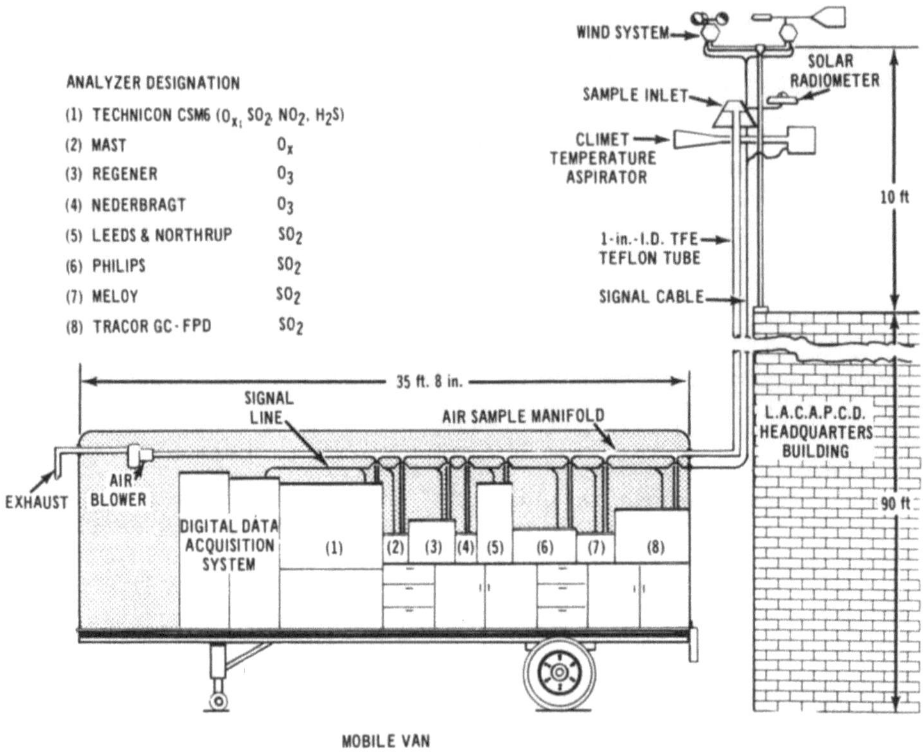

Figure 1. Diagram of Air Sampling System and Sulfur Dioxide and Ozone Monitors in Mobile Laboratory.

TABLE I

Sulfur Dioxide Instruments and Detection Principles

Manufacturer	Model	Detector Principle
Technicon Instruments	CSM-6	West-Gaeke Colorimetric
Leeds & Northrup Corp. North Wales, Pennsylvania	Aeroscan-7860	Conductometric
Philips Electronics Mount Vernon, New York	PW-9700	Coulometric
Meloy Laboratories Springfield, Virginia	LL-1100	Flame Emission
Tracor Incorporated Austin, Texas	250 Series	Gas Chromatographic Flame Emission Combination

EXPERIMENTAL*

Data Acquisition System -- As part of the basic requirements of the instrument evaluation program, data had to be recorded at 5-minute intervals in order to facilitate the short term instrument evaluation. Since the evaluation program included ten pollution monitors plus meteorological instruments, output signals had to be obtained and recorded automatically in digital form. The data acquisition system used in this study consisted of signal-conditioning circuitry, on-line digital and analog recording systems, and power supply units. A functional diagram of the data acquisition system is shown in Figure 2.

The digital recording system (Dymec Model 2015H) consists, basically, of a crossbar scanner (Hewlett-Packard Model 2911) for connecting each individual channel in sequence to the input of the analog-to-digital converter (A/D), which in this system is a digital voltmeter (Hewlett-Packard Model 3460A). The 12-character data word from the A/D converter--which includes six characters for the data point, two characters for channel identification, plus polarity and range information--is gated into the format control unit (DY Model 2546B). This unit then feeds the data in serial form into the digital incremental magnetic tape recorder (Kennedy 1400) and, in parallel, to the digital printer (Hewlett-Packard 562A). The data acquisition system used has an 8-inch reel of 1.5 mil tape, a 200-bit-per-inch packing density, and a 33-word record, and samples 65 channels each 5-minute period. Consequently, the system can record continuously for approximately 8.4 days on one reel of tape. Tape changes which require about 5 to 10 minutes, can be made weekly, so that few data are lost. The digital printer is coupled in parallel with the magnetic tape recorder and is used for monitoring the digital data as required.

The data acquisition system was designed to provide accurate, reliable, computer-compatible data, and has provisions for introducing status information and calibration data for adding or deleting instruments without disturbing the continuous recording of digital data.

Calibration of Sulfur Dioxide Instruments -- A dynamic calibration system, using a gravimetrically calibrated SO_2 permeation tube[4] as a primary standard and zero air as diluent, was used to provide known concentrations of SO_2 to the respective analyzers. Figure 3 shows the permeation tube calibration system used for this investigation. The permeation tube was housed in a Pyrex holder and immersed in a constant temperature bath that maintained the tube at a temperature of 20.3° ± 0.1°C.

*Mention of a company or product name does not constitute endorsement by the U. S. Environmental Protection Agency.

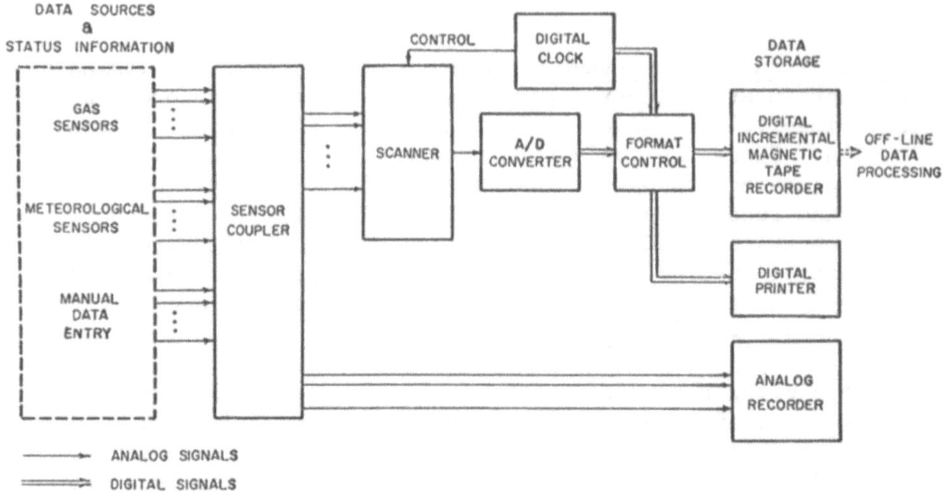

Figure 2. Functional Diagram of Data Acquisition System.

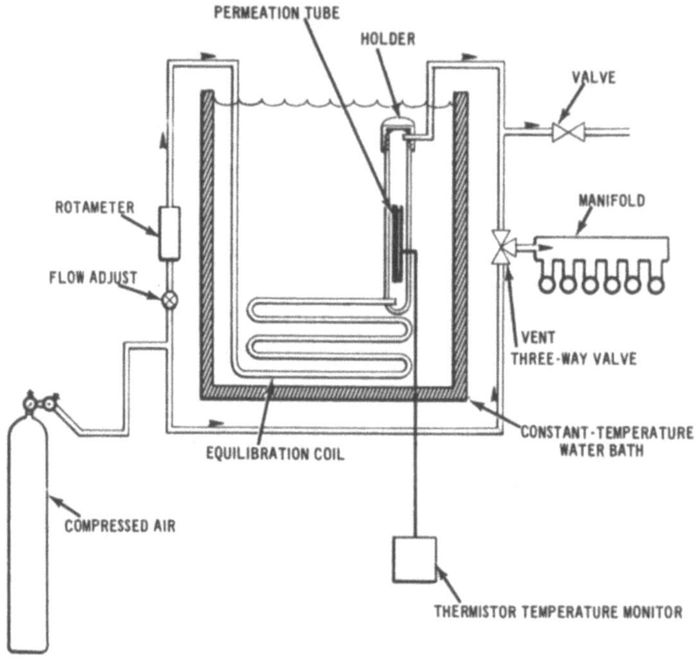

Figure 3. Permeation Tube Calibration System.

Dry compressed air, conditioned to the temperature of the bath and metered through a rotameter, was passed over the permeation tube and into a 1 inch (O.D.) glass manifold that was installed behind the SO_2 analyzers. Ball-and-socket connections were used to allow easy hookup of the instrument sample inlet lines to the manifold during calibration. By varying the diluent air flow rate, SO_2 concentrations of 0 to 0.2 ppm were generated and simultaneously supplied to all the SO_2 monitors during calibration. Calibration gas in excess of instrument requirements was maintained at all times. The permeation tube used during this field evaluation program was calibrated by the National Bureau of Standards.

Definition of Terms -- In this manuscript, several terms are used to describe the response characteristics of the instruments: lag time, rise time, fall time, minimum detectable concentration, and analysis range. These terms are defined below:

Lag time (Initial response time): The time interval from a step change in input concentration at the instrument to the first corresponding change in the instrument output.
Rise time: Interval between the initial response time and the time to 95 percent response after a step increase in input concentration.
Fall Time: Interval between the initial response time and the time to 95 percent response after a step decrease in input concentration. This time is not necessarily equal to the rise time but may be approximately the same.
Minimum detectable concentration: The smallest amount of input concentration that can be detected with a specified degree of confidence.
Analysis range: The minimum and maximum measurement limits. The effective range may be limited by the points where no readable response can be obtained.

Operational Characteristics of Sulfur Dioxide Instruments -- For this study, response characteristics of each monitor were determined by dynamically introducing known concentrations of sulfur dioxide to each sensor. Table II lists the results of this study. Response-time factors for the gas chromatographic techniques are not applicable because the duration of the analysis is the same as the chromatographic-cycle period (in this case, one sample is injected every 5 minutes). Sulfur dioxide is eluted after 3 minutes, which is the actual time required for analysis by the Tracor-GC-FPD analyzer. The chromatographic analyzer[2] also simultaneously measures H_2S so that for every air sample injected, two measurements are performed by the Tracor-GC-FPD.

TABLE II

Instrument Response Characteristics

	Performance Factors				
Instrument	Analysis Range[1] ppm	Minimum Detectable, ppm	Lag Time[2]	Rise Time to 95%	Fall Time to 95%
Technicon-West Gaeke	0-0.3	0.010	25.0	9.0	9.0
Leeds & Northrup-Conductivity	0-0.3	0.010[3]	1.0	5.0	5.0
Philips - Coulometric	0-1.0	0.015	2.0	1.0	1.0
Meloy-FPD	0-0.3	0.005	0.5	1.0	1.0
Tracor-GC-FPD	0-0.8[4]	0.005	3.0	NA[5]	NA

(1) Range set for this study.

(2) Time response reported in minutes

(3) Signal amplified to give increased sensitivity not standard with instrument.

(4) Range to data acquisition system was 0-0.8 ppm; recorder range 0-0.3 ppm.

(5) NA = Not Applicable.

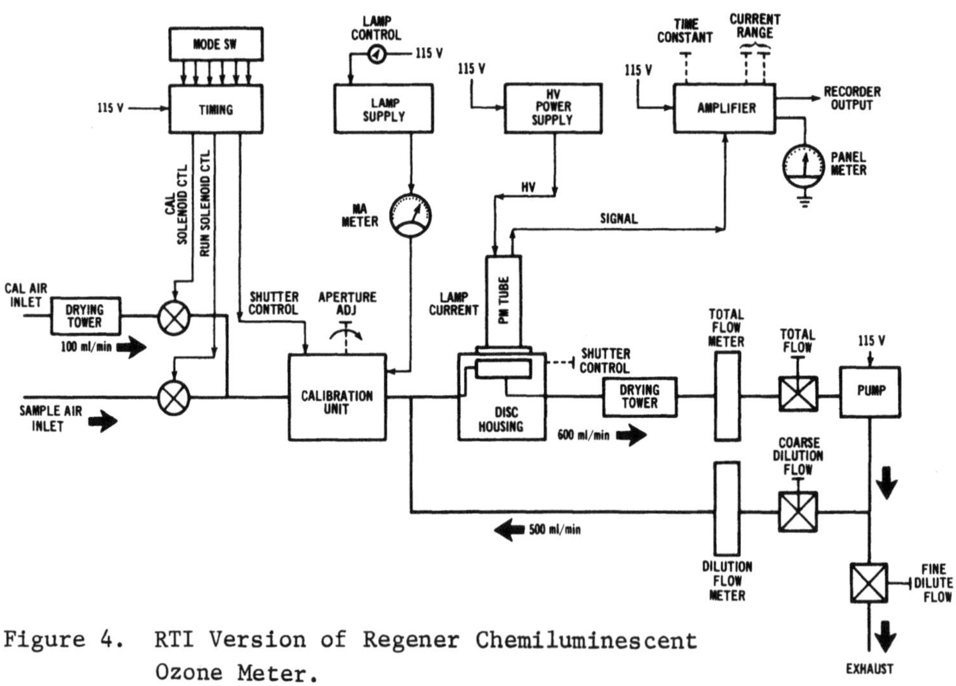

Figure 4. RTI Version of Regener Chemiluminescent Ozone Meter.

Operational Characteristics of Ozone and Oxidant Analyzers -- Two oxidant analyzers based on the reaction of an oxidizing substance with potassium iodide were used in this study. One of these analyzers was a coulometric unit, employing KI/KBr reagent. This analyzer was manufactured by the Mast Company, Davenport, Iowa.

The other oxidant analyzer was supplied by Technicon Corporation and was included as one of the systems in Technicon's CSM-6 console described above. This analyzer utilized continuous colorimetric measurement of iodine produced in 10 percent KI reagent. This unit is characteristic of analyzers being commonly used today by state and local air pollution agencies.

Two detectors employing different chemiluminescent reactions of ozone were used for specific measurements of ozone. An automated instrument based on Regener's principle was constructed and supplied by Research Triangle Institute, Research Triangle Park, N. C., under an Environmental Protection Agency Contract and has been described previously[5]. The Regener system operates by measuring the chemiluminescence that results from ozone passing over silica gel coated with Rhodamine-B. The intensity of the luminescence is proportional to the ozone concentration. A schematic of that detector is shown in Figure 4.

The second ozone detector was a laboratory prototype unit similar to the systems described by Nederbragt[6] and Warren[7]. The Nederbragt detector measures ozone by means of the chemiluminescent reaction between ozone in air with ethylene. A schematic of this detector is shown in Figure 5. Air flowing at 1 liter per minute mixes with pure ethylene at 30 ml/minute at the tip of two concentric tubes and impinges on a pyrex window closely coupled to the photocathode. At atmospheric pressure, the chemiluminescent intensity is proportional to ozone concentration. The response characteristics of this detector and a comparison with the Regener detector have been reported previously[8].

DISCUSSION OF RESULTS

Sulfur Dioxide Study

Calibration Stability of SO_2 Monitors -- Ideally an air pollution monitor should experience less than 1 percent calibration drift at full scale of the measuring range over a 1 week period. Unfortunately, most air monitors usually drift considerably more than this. An important part of this program was to determine the extent of calibration drift for each instrument over the measuring range of interest. For Los Angeles, the mean concentration of SO_2

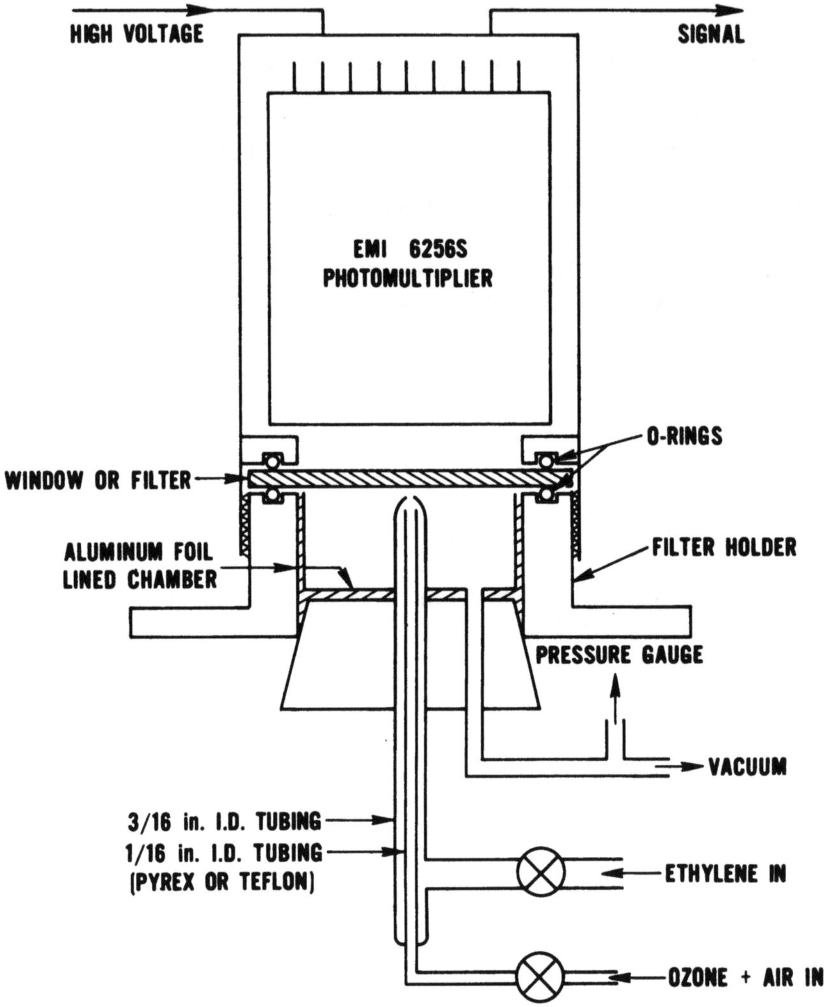

Figure 5. Nederbragt Chemiluminescent Ozone Detector.

observed between September and December 1970 was less than 0.02 ppm. A drift of 1 percent per 24 hours on a sensitivity setting of 0.5 ppm full scale would, therefore, produce a standard deviation of 25 percent in the mean SO_2 concentration and adversely affect the precision and accuracy of the measurements.

To determine the extent of the calibration drift, a multiple-point calibration for each instrument was performed. All instruments were calibrated dynamically by means of an SO_2 permeation tube, calibrated by the National Bureau of Standards about once every 2 weeks over the 3-month study period. Calibration data, summarized in Table III, show the standard deviations of the slope and intercept between calibration periods. The standard deviation of the slope of the calibration curve for the Technicon-West-Gaeke was 26 percent. This value would have been even larger if zero adjustments, necessitated by a continuous signal drift upscale, had not been made almost daily to this instrument. Some of this drift may have resulted from exposure of the dyes to ultraviolet radiation emanating from the fluorescent lamps mounted in the trailer.

Table IV lists the observed standard deviation, in ppm, obtained from the least squares plot of the calibration data. The Technicon-West-Gaeke and Meloy-FPD deviated less than 0.007 ppm, whereas the Philips coulometric and Leeds & Northrup conductometric analyzers showed significant calibration variations at 0.04 ppm. These instruments were marginally acceptable even at 0.08 ppm. The Tracor-GC-FPD system calibration drift was within acceptable limits. The time required to prepare a four-point calibration curve for each instrument is shown in Table V.

Although all the SO_2 instruments except the Technicon-West Gaeke required less than 1.25 hours for calibration, the long response time of the Technicon affected the amount of valid data collected by the other instruments. Because the format of the calibration program was set up to deliver known concentrations of SO_2 to each monitor simultaneously from a common manifold, the time required for each calibration point was limited by the instrument with the slowest response.

Statistical Comparison of Instruments

Frequency Distribution -- The mean SO_2 concentration detected by the monitors over the 83-day comparative study was below 0.02 ppm. At these concentrations, casual computations attempting to correlate responses of the instruments can be misleading.
Table VI lists the frequency distribution of SO_2 concentrations for each instrument.

TABLE III

Summary of All Calibration Data

Technicon-West-Gaeke	m	b		N[1]
Average	0.0241	- 0.0068		23
Standard Deviation	0.0062	0.0084		
Philips-Coulometric	m	b		N
Average	18.561	-0.1016		24
Standard Deviation	2.691	0.0238		
Leeds & Northrup-Conductometric	m	b		N
Average	0.3000	-0.0010		20
Standard Deviation	0.0918	0.00371		
Meloy-FPD	a	n	I_o	N
Average	0.137	0.654	0.0068	12
Standard Deviation	0.0164	0.0526	0.0040	
Tracor-GC-FPD	a	n	I_o	N
Average	0.02822	0.6533	--	9
Standard Deviation	0.00335	0.02286	--	

Calibration equations for Technicon, Philips, Leeds & Northrup[2]

$y = mI + b$ where
y = concentration, ppm
m = slope
b = zero intercept
I = signal, mV

Calibration equations for Meloy and Tracor

$y = a(I - I_o)^n$
y = concentration, ppm
a = coefficient
n = slope (log-log plot)
I = signal, mV
I_o = baseline signal offset, mV

(1) N = Number of Calibrations
(2) Calibration equation for Technicon-West-Gaeke, Philips-Coulometric; Leeds & Northrup Analyzer, Linear Slope

TABLE IV

Standard Deviation in Calibration Data at Several Concentrations

$$\sigma = \left(\frac{\Sigma d^2}{N-1}\right)^{1/2\,(a)}$$

	Sulfur Dioxide Sensors				
Concentration, ppm	Technicon Colorimetric West-Gaeke σ, ppm	Meloy FPD σ,ppm	Philips Coulometric σ, ppm	Leeds & Northrup Conductometric σ, ppm	Tracor GC-FPD σ, ppm
0.000	0.007	0.005	0.012	0.001	0.000
0.021	0.005	0.004	0.010	0.007	0.004
0.040	0.003	0.004	0.014	0.014	0.006
0.080	0.003	0.005	0.013	0.014	0.010

[a] σ = Standard deviation, ppm
d^2 = Individual variations
N = Number of measurements

TABLE V

Time Required to Perform 4-Point[(1)] Calibration Curve for SO_2 Monitors.

Instrument	Time, hours
Technicon-West-Gaeke	3
Philips-Coulometric	1
Leeds & Northrup-Conductometric	1
Meloy-FPD	1
Tracor-FPD-GC	1.25

[(1)] Calibration includes zero and three sulfur dioxide concentrations.

TABLE VI

Frequency Distribution of Sulfur Dioxide Measurements

Concentration ppm	Technicon-West-Gaeke	Meloy FPD	Philips Coulometric	Leeds & Northrup Conductometric	Tracor GC-FPD
	Percent of Measurements[1] Exceeding Concentration in Column 1				
0.000	89.3	81.4	80.0	93.8	95.4
0.003	71.7	74.5	84.5	88.7	87.4
0.005	69.3	60.6	68.2	72.0	70.6
0.010	55.5	42.4	55.8	57.3	43.2
0.015	43.6	16.9	43.9	48.1	20.6
0.020	33.0	8.8	33.8	38.6	10.5
0.030	21.0	2.5	19.7	22.5	3.7
0.050	7.4	0.5	5.8	8.7	0.5
0.075	1.7	0.1	0.6	3.1	0.1
0.100	0.5	0.1	0.2	0.7	0.1
0.150	0.1	< 0.1	0.1	0.2	< 0.1
0.200	< 0.1	< 0.1	< 0.1	0.1	< 0.1

[1] Measurements made at 5-minute intervals.

Only 3 times during the entire monitoring period did the SO_2 concentration for all instruments exceed 0.100 ppm, and the duration of these events was only 15 to 20 minutes. As indicated in Table IV, the relative standard deviation of the calibration data, even at 0.04 ppm, ranged from 9 to 35 percent. So that the performance of these analyzers could be examined and objectively compared data were selected by the following criteria: at least four of the five instruments must be in operation and the concentration of SO_2 must be above 0.030 ppm.

Correlation of Data -- Table VII is a correlation matrix of 24 hourly averages selected over the 83-day monitoring period. The hourly average represented 12 measurements, one taken every 5 minutes. Correlation above 0.70 indicates that instrument responses generally followed the same pattern but did not necessarily record the same concentrations. Mean values, along with correlation coefficients, can be used to compare measurements on an absolute basis. For example, the Technicon-West-Gaeke and Tracor-GC-FPD had a correlation of 0.87 and mean concentrations, respectively, of 0.036 and 0.040 ppm. The correlation between the Leeds & Northrup conductivity monitor and the Tracor-GC-FPD was only 0.139 and the Leeds & Northrup conductivity mean average was 0.063 ppm, twice the Tracor-GC-FPD value. The conclusions based on these computations were: (1) the Leeds & Northrup conductivity and Philips coulometric analyzers exhibited poor correlations with the Technicon-West-Gaeke, Tracor-GC-FPD and Meloy-FPD, and (2) the Philips coulometric and Leeds & Northrup conductometric analyzers record SO_2 concentrations as much as twice the values measured by the other monitoring principles. Almost every instance in which the recorded SO_2 concentration exceeded 0.050 ppm, occurred between 1200 and 1600 hours. Substances such as fine particulate matter, ozone, oxides of nitrogen, and certain organics, which are generated during the process between 0880 and 1600 by industrial and automotive activity, can find their way into the coulometric or conductometric cells and produce positive interference. Such interference might explain the consistently higher SO_2 concentrations recorded by the Philips and Leeds & Northrup analyzers.

Tables VIII and IX are correlation matrices of data taken for 2 of the 3 days the SO_2 in downtown Los Angeles exceeded 0.100 ppm. Figures 6 and 7 present variations in SO_2 concentrations with time for these 2 days (October 30, 1970, and November 24, 1970). All instruments had reasonable correlations on both days. On October 30, 1970, the mean concentration for the Tracor-GC-FPD was 0.068 ppm, with a peak value of 0.125 ppm, while the mean value for the Leeds & Northrup conductivity monitor was 0.110 ppm, with a peak value of 0.183 ppm. On November 24, 1970, the Philips coulometric monitor recorded a peak SO_2 concentration at 1320 hours of 0.222 ppm, while the Meloy-FPD, Tracor-GC-FPD, and the Technicon-

TABLE VII

Sulfur Dioxide Correlation Coefficients and Mean Concentrations for Measurements Made Between September 11 and November 24, 1970[1]

Instrument	Technicon-Colorimetric	Meloy FPD	Philips Coulometric	Leeds & Northrup Conductometric	Tracor GC-FPD
Technicon Colorimetric West-Gaeke	1				
Meloy FPD	0.805	1			
Philips Coulometric	0.721	0.810	1		
Leeds & Northrup Conductometric	0.140	0.663	0.572	1	
Tracor GC-FPD	0.870	0.743	0.432	0.139	1
Mean Concentration, ppm	0.036	0.033	0.050	0.063	0.040

[1] Data selected for this correlation matrix were restricted to hourly average SO_2 concentrations exceeding 0.02 ppm.

TABLE VIII

Sulfur Dioxide Correlation Coefficients and Mean Concentrations for Measurements Made October 30, 1970, 1200 to 1400 Hours[1]

Instrument	Technicon Colorimetric	Meloy FPD	Philips Coulometric	Leeds & Northrup Conductometric	Tracor GC-FPD
Technicon Colorimetric West-Gaeke	1				
Meloy FPD	0.882	1			
Philips Coulometric	0.907	0.995	1		
Leeds & Northrup Conductometric	0.933	0.983	0.994	1	
Tracor GC-FPD	0.952	0.826	0.847	0.880	1
Mean Concentration, ppm	0.104	0.089	0.104	0.110	0.068

[1] 5-Minute Intervals.

TABLE IX

Sulfur Dioxide Correlation Coefficients and Mean Concentrations for Measurements Made October 24, 1970, 1200 to 1355 Hours[(1)]

Instrument	Technicon[(2)] Colorimetric	Meloy FPD	Philips Coulometric	Leeds & Northrup Conductometric	Tracor GC-FPD
Technicon Colorimetric West-Gaeke	---				
Meloy FPD	---	1			
Philips Coulometric	---	0.967	1		
Leeds & Northrup Conductometric	---	0.978	0.962	1	
Tracor GC-FPD	---	0.887	0.309	0.927	1
Mean Concentration, ppm	---	0.084	0.177	0.118	0.082

(1) 5-Minute Intervals

(2) Colorimetric data not available.

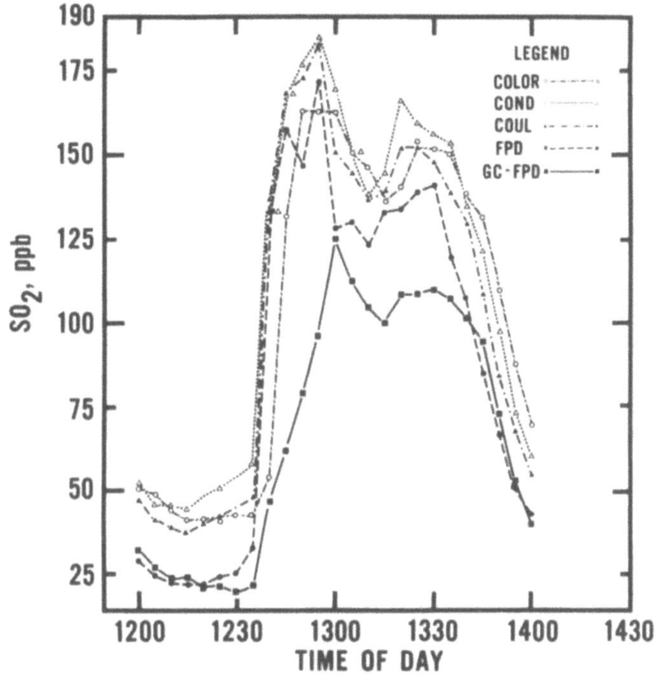

Figure 6. Sulfur Dioxide Concentration Between 1200 and 1500 Hours on October 30, 1970.

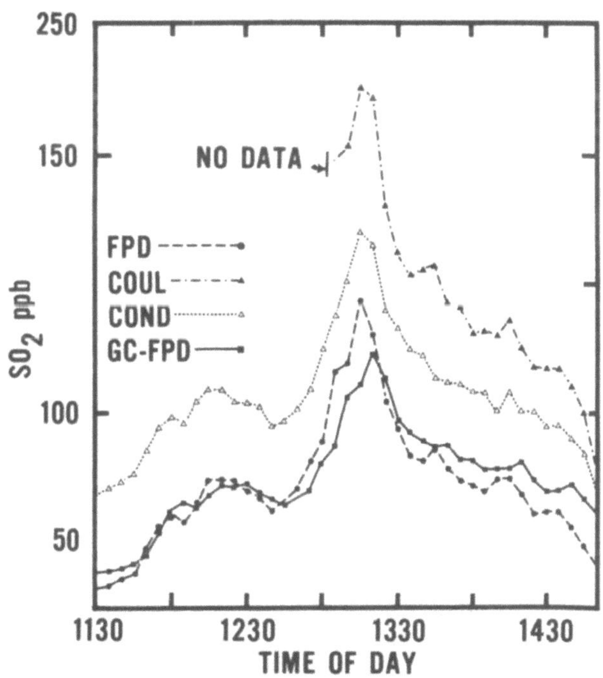

Figure 7. Sulfur Dioxide Concentrations Between 1200 and 1500 Hours on November 24, 1970.

West-Gaeke recorded values between 0.123 and 0.147 ppm. On this day, ozone and oxidant concentrations did not exceed 0.030 ppm. The nature of the positive interferences with the Philips coulometric analyzer on this day is probably not related solely to ozone but to some other species indigenous to the area. The manufacturer lists a number of substances that potentially interfere with coulometric measurements for SO_2. Among these substances are unsaturated hydrocarbons and aldehydes, both of which are known to be present in the Los Angeles atmosphere.

Maintenance -- During the evaluation period, the time required to keep each monitor in continuous operation was carefully recorded. Table X summarizes calibration, maintenance, and downtime for each SO_2 monitor. Two of the instruments required the addition of reagents periodically for the study. The time required to replenish these reagents was considered a maintenance operation and is listed in Table X. The Meloy-FPD and the Tracor-FPD-GC require replacement of gas cylinders, but this operation requires less than 10 minutes, so that the instruments are out of service for only 30 minutes at most. Consequently, the Meloy-FPD and Tracor-GC-FPD had the least downtime of all the monitors evaluated.

The major maintenance problems encountered with the other monitors were as follows: (1) Technicon-West-Gaeke: the logarithmic amplifier failed and was replaced; pump tubing was replaced biweekly which necessitated recalibration of the monitor each time; (2) Leeds & Northrup conductivity: the reagent pump lost prime because of air bubbles; and (3) Philips coulometric: one electrode became defective and was replaced; fuse failure occurred.

Ratio of Total Gaseous Sulfur to SO_2 -- The Meloy-FPD and the Tracor-GC-FPD measurements for this study generally agreed, except for two instances. On September 8, 1970 for 15 minutes, the Meloy-FPD recorded values up to 0.160 ppm, while the gas chromatographic-FPD analyzer recorded concentrations of 0.050 ppm for H_2S, 0.025 ppm for SO_2, and 0.050 ppm for CH_3SH. On October 30, 1970, the Tracor-GC-FPD recorded a peak concentration of SO_2 of 0.129 ppm, while the Meloy-FPD recorded a 0.173 ppm peak. This discrepancy of 0.044 ppm could not be explained by the presence of H_2S because the chromatographic value for H_2S during this period was less than 0.01 ppm. If we can assume that these two instances were of short enough duration to be insignificant, we can conclude that the total gaseous sulfur concentration in downtown Los Angeles is better than 90 percent sulfur dioxide.

One channel of the Technicon-CSM-6 was equipped to measure H_2S by the methylene-blue procedure described by Jacobs, Braverman, and Hochheiser[9]. The analytical data obtained with this method are inconclusive because of the continuous baseline drift of the signal.

TABLE X

Maintenance and Downtime Summary

Instrument	Calibration	Maintenance	Awaiting Maintenance	Repairs	Misc. Downtime	Total Off-line	Valid Data Collection
Technicon-West Gaeke	4.0	3.1	1.9	4.5	12.1 (1)	25.6	74.4
Leeds & Northrup-Conductometric	2.9	0	0	1.1	4.1	8.1	91.9
Philips-Coulometric	2.7	0 (3)	0	0.04	7.7	10.4	89.6
Meloy-FPD	3.0	0.13 (2)	0	0.09	2.9	6.1	93.9
Tracor-FPD	2.9	0.09	0	0	4.1	7.1	92.9

(1) All values expressed as percent of time for 90-day evaluation period.

(2) Log-amplifier failure.

(3) Maintenance recommended at 90-day intervals (3-hours required).

CONCLUSIONS

One of the major objectives of this study was to compare performance characteristics of new analytical measurement techniques for SO_2 with classic procedures. In addition, we hoped to be able to determine the ratio of SO_2 to total gaseous sulfur in Los Angeles, California. The sulfur dioxide concentration, however, in Los Angeles averaged 0.02 ppm or below 70 percent of the time. Thus, relatively few measurements were available on which to base a completely satisfactory correlation study.

Nevertheless, the following conclusions and important observations can be cited based on the data that were available and considered to be valid:

1. Coulometric and conductometric SO_2 measurements are generally higher than values obtained with the colorimetric-West Gaeke, flame photometric, and chromatographic-FPD analyzers.
2. The flame photometric, gas chromatographic-FPD, coulometric, and conductometric analyzers required the least maintenance and were operational for more than 93 percent of the evaluation period.
3. Hydrogen sulfide and methyl mercaptan are occasionally present in the Los Angeles atmosphere.
4. Sulfur dioxide frequently exceeds 0.030 ppm during the afternoon.
5. Mean concentration of SO_2 was less than 0.020 ppm during the evaluation period.
6. Total gaseous sulfur in Los Angeles, California was better than 90 percent SO_2.

Ozone-Oxidant Study

Calibration Stability -- The ozone and oxidant instruments were calibrated daily during the first two weeks of the study. Later the frequency of calibration was reduced to a biweekly basis. Usually 4 to 6 calibration points between 0 and 0.5 ppm ozone were obtained. The calibration curves for each of the instruments were linear over this range.

Table XI shows the reduced calibration data. The frequency of calibration is not the same since all instruments were not operational for the entire time. The Nederbragt detector showed the smallest standard deviation of slope giving a value of 2.1 percent. The 14 percent standard deviation of the colorimetric oxidant analyzer represents the poorest stability observed.

TABLE XI

Reduced Calibration Data

Instrument	$\bar{m}$, ppm/volt[a]	$\bar{b}$, ppm[a]	$\sigma(\bar{m})$[b]	$\sigma(\bar{b})$[b]	N[c]
Regener[d]	0.107	--	0.0082	--	18
Nederbragt	0.475	-0.0035	0.010	0.0027	7
Technicon	0.0922	-0.0032	0.0132	0.0115	18
Mast 1	52.62	-0.0002	2.27	0.001	14
Mast 2	66.96	-0.001	4.32	0.0016	8

a) $[O_3] = \bar{m} \times V + \bar{b}$; $[O_3]$ = concn, ppm
V = response, volts or mV

b) Standard Deviations in Slope, $\bar{m}$, and Intercept, $\bar{I}$

c) N = Number of Calibration Curves

d) Data reduced from strip chart recorder.

Sulfur Dioxide Interference -- Sulfur dioxide when present in lower concentration than ozone produces a quantitative negative interference on the oxidant instruments. This effect was noted on several occasions in Los Angeles during this study. Figure 8 shows clearly the SO_2 interference on the Mast meter for one day (10/21/70) during which sulfur dioxide concentrations reached particularly high levels. Pronounced minima were observed in Mast readings at 1230 and 1430 hours. These periods corresponded closely with the sulfur dioxide maxima.

Correlation Between Ozone and Oxidants -- During the 3-month study, all the daylight data for the ozone and oxidant measurements were analyzed to determine the relationship between oxidants and ozone. One of the objectives of the study was to determine if total oxidant values, corrected for the NO_2 interference during the daylight maxima, are essentially equivalent to the ozone concentration. If this relationship could be substantiated then urban centers can use ozone monitors as one measure of their air quality rather than oxidant analyzers. This would be a preferred procedure since chemiluminescent ozone analyzers are virtually free from interferences and considerably easier to operate and maintain than current state-of-the-art oxidant analyzers.

Figure 9 shows the diurnal averages between September 4, 1970 and September 30, 1970 for the colorimetric and coulometric oxidant analyzers and the Nederbragt chemiluminescent ozone analyzer. The oxidant analyzer measurements were corrected for NO_2. These data indicate that the corrected oxidant values are essentially equivalent to the ozone measurements while the coulometric measurements were about 24 percent lower than ozone values at the maxima.

CONCLUSIONS

1. The correlation between ozone concentrations obtained by chemiluminescent methods and oxidants was 70 percent for daytime hours.
2. The comparison between ozone and oxidant concentrations corrected for sulfur dioxide and nitrogen dioxide interferences exhibited poor correlation. These results were probably mainly due to uncertainties in the NO_2 determinations.
3. The mean ozone values were approximately 98 percent of the mean oxidants for the three-hour interval of maximum oxidants concentrations in downtown Los Angeles, (1100 - 1400 hours).
4. Mean ozone was equal to 126 percent of the coulometric oxidant average over the same interval.

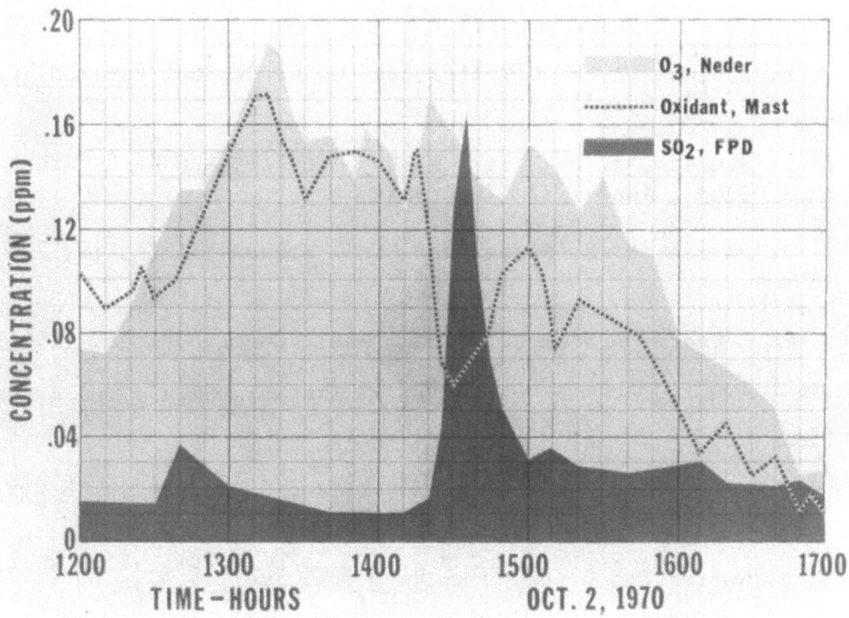

Figure 8. Effect of Sulfur Dioxide Incident as Oxidant Monitor.

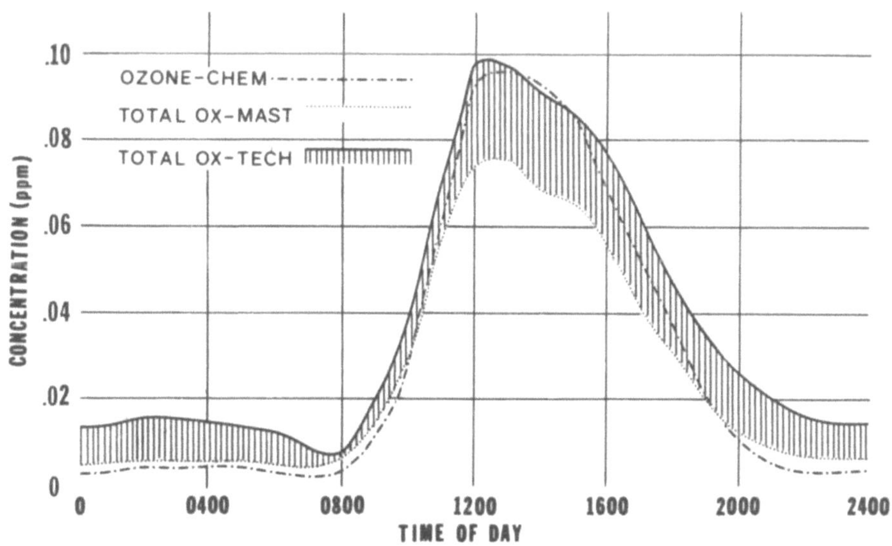

Figure 9. Diurnal Ozone-Oxidant Averages Between September 4 through September 30, 1971.

REFERENCES

1. R. K. Stevens, A. E. O'Keeffe, and G. C. Ortman, Environ. Sci. Technol., 3, 652 (1969).

2. R. K. Stevens, J. D. Mulik, A. E. O'Keeffe, and K. R. Krost, Anal. Chem., June 1971 (In Press).

3. H. F. Palmer, C. E. Rodes, and C. J. Nelson, JAPCA, 19, 778 (1969).

4. A. E. O'Keeffe and G. C. Ortman, Anal. Chem., 38, 760 (1966).

5. J. A. Hodgeson, K. R. Krost, A. E. O'Keeffe, and R. K. Stevens, Anal. Chem., 42, 1795 (1970).

6. G. W. Nederbragt, A. Van der Horst, and J. Van Duijn, Nature, 206, 87 (1965).

7. G. J. Warren and G. Babcock, Rev. Sci. Instr., 41, 280 (1970).

8. J. A. Hodgeson, B. E. Martin, and R. E. Baumgardner, Prog. Anal. Chem., 5, Plenum Press (1971).

9. M. B. Jacobs, M. M. Braverman, and S. Hochheiser, Anal. Chem., 29, 1349 (1957).

ATMOSPHERIC OZONE DETERMINATION BY AMPEROMETRY AND COLORIMETRY

Y. Tokiwa, SuzAnne Twiss, E. R. de Vera and P. K. Mueller

Air and Industrial Hygiene Laboratory

California State Department of Public Health

INTRODUCTION

The iodometric procedure has been used extensively to monitor oxidant and ozone concentrations in atmospheric reaction studies, chamber exposures and calibration of oxidant and ozone analyzers as well as to collect air quality data. The term "oxidant" refers to all substances present in the sampled air that oxidize potassium iodide (KI) to form iodine (I_2) such as ozone (O_3), nitrogen dioxide (NO_2), and organic peroxides. Specific spectroscopic measurements indicate that ozone is the principal oxidant in most atmospheres.

Continuous analysis of oxidant has most commonly been performed by scrubbing sample air with an absorbing solution of potassium iodide (KI) buffered at pH 6.8. Oxidants react with the KI to form triiodide ion (I_3^-). The concentration of I_3^- is measured by a colorimeter, amperometric cell, or potentiometric cell and recorded as ppm oxidant. Since a standard "oxidant" cannot be prepared, oxidant analyzers are calibrated with mixtures of ozone in filtered air.

The iodometric procedure, however, is not specific. While the response of the KI reagent to ozone is 100%, variable amounts of iodine are released by nitrogen dioxide (NO_2), peroxyacetyl nitrate (PAN) and other oxidizing substances. Sulfur dioxide (SO_2) decreases the oxidant readings by 100% of the SO_2 present. The levels of PAN and other oxidizing substances are generally considered low (0.001 to 0.1 ppm) and therefore do not significantly interfere. The response to NO_2 varies with the reagent formulation, contact column design and measurement principle (1-5). In California, 2, 10 and 20% neutral buffered KI solutions are used by various monitoring agencies.

More recently, improved techniques for detection of ozone specifically have been developed. These methods measure ozone in the gas phase by detecting either the absorption of ultraviolet radiation by the ozone molecule or by the production of chemiluminescence when ozone is reacted with a specific gas, usually ethylene, or on a suitable surface. Other pollutants normally found together with oxidant and ozone do not interfere (5).

The major interferences in the iodometric procedure are, therefore, due to NO_2 and SO_2. Various filters and scrubbers for removing SO_2 have been suggested and used, but detailed information concerning their performance is not available. One apparently effective filter consists of glass fiber strips impregnated with chromium trioxide (CrO_3) and sulfuric acid (H_2SO_4). However, this filter is equally effective in converting nitric oxide (NO) to NO_2. Therefore, in atmospheres containing significant amounts of NO, the use of this filter is not recommended (5).

This lack of specificity and dependence on instrument design indicates that oxidant data collected by different iodometric methods are not entirely comparable. However, such oxidant data can be compared provided the following information is available: 1) the identity of the analyzer 2) the reagent formulation 3) the concurrent concentrations of all interfering substances in the sampled air, 4) the magnitude of the signal produced by each interference and 5) any user modifications which may affect item 4. This information may then be used to correct air monitoring data and permit comparison with data collected by others. In addition, information concerning maintenance practices and procedures and frequency of calibrations will materially assist in interpretation of such data.

Reported here are the results of our studies concerning the effect of NO_2 on different iodometric procedures for oxidants, and utilizing this information to correct and compare actual air monitoring data.

METHODOLOGY

The investigations were conducted in two phases: Phase I established the response of different iodometric oxidant analyzers to NO_2 and Phase II verified the validity of correcting air monitoring data for the effect of NO_2. The instruments utilized in this study are listed in Table I. The scrubbing columns are specified because their design governs both the transfer efficiency of ozone and of other interfering gases. Both reagent composition and contactor configuration affect the interference equivalents.

All analyzers were initially calibrated with mixtures of ozone or NO_2 in filtered air (9-12) and recalibrated periodically throughout the study period. Filtered air was obtained by passing ambient air through a cartridge containing activated charcoal and soda lime (13). The cartridge, designed to remove organic vapors and acid gases, efficiently removes ozone, NO and NO_2. The accuracy of all instrument air and liquid flow metering devices was also verified at periodic intervals.

Phase I was conducted in the laboratory with three oxidant analyzers; an amperometric unit using 2% KI (6) and two colorimetric units (7), one using 10% KI and the other using 20% KI. The magnitude of the response to NO_2(interference

Table I

OXIDANT ANALYZER CHARACTERISTICS

			CONTACTOR			AIR LIQUID FLOW		
ANALYZER	REF. NO	REAGENT	Configuration	Length (cm)	Direction	Air l/min	Liquid ml/min	Ratio l/ml
Amperometric	6	2% KI† 5% KBr	wire helix on central support	5	concurrent down	0.140	0.042	3.3
Colorimetric	7	10% KI††						
Phase I			helix insert	50	counter-current	4.0	4.0	1.0
Phase II			helix insert	23	counter-current	2.0	2.0	1.0
Colorimetric	7	20% KI††	2-1/2 turn helical coil	84	concurrent up	2.0	2.0	1.0

† reagent buffered at pH 6.8 ±0.01 with 0.026 M Na_2HPO_4 and 0.018 M NaH_2PO_4

†† reagent buffered at pH 6.8 ±0.01 with 0.1 M Na_2HPO_4 and 0.1 M KH_2PO_4

equivalent) was established for each of the three analyzers utilizing mixtures of NO_2 and O_3 in filtered air in the range of 0.0 to 60 pphm for each gas.

Data for Phase II was collected under actual air monitoring conditions by the amperometric unit and a modified colorimetric 10% KI analyzer. Concurrent NO_2 concentrations were monitored with a colorimetric NO_2 analyzer (8) utilizing a common sampling manifold.

The amperometric and colorimetric data corrected for NO_2 interference were then compared. When the data are the same, the validity of the interference equivalents would be verified. When data from the two analyzers are significantly different then the following possibilities would be evident: 1) the interference equivalents may be incorrect, i.e., the ratio of the NO_2/O_3 concentrations in the ambient air may have been outside the range of the tests conducted in the laboratory, or 2) interferences in addition to NO_2 were present.

Amperometric Oxidant Analyzer

The operational principles of the amperometric unit are described in the Intersociety Committee's manual of methods for ambient air sampling and analysis (14). The contactor-sensor for this unit consists of a plastic block with a hole about 0.6 cm ID and 5 cm long, in which a rod, wound with many turns of a fine platinum wire (cathode) and two turns of a heavier wire (anode), is located axially (See Figure 1).

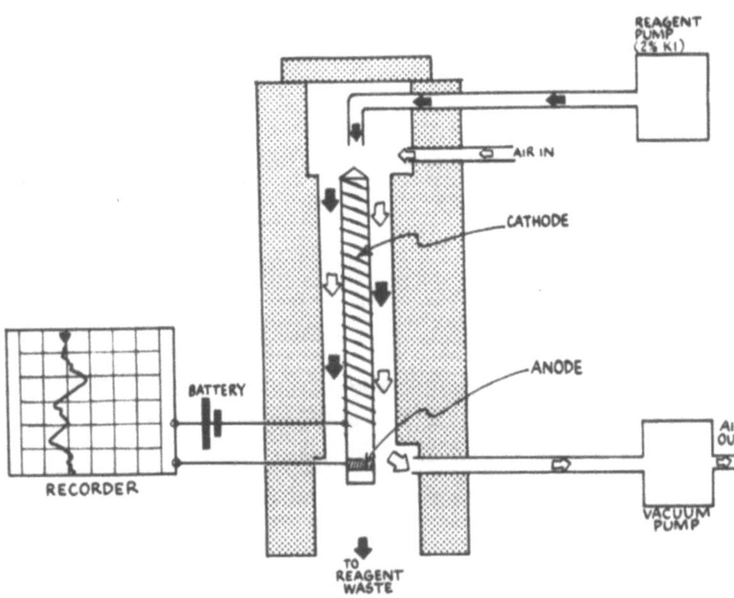

Figure 1. OXIDANT ANALYZER - AMPEROMETRIC

Sample air and reagent pass concurrently through the annular space (about 0.2 cm) between the support rod and the cylinder wall. Ozone in air transferred to the reagent reacts with the iodide to produce I_3^- which in turn reacts with the hydrogen polarizing the cathode. As a result of these reactions, current flows to repolarize the cathode in proportion to the amount of hydrogen removed. This current, which is directly proportional to concentration of ozone, was recorded on a 4-inch wide rectangular coordinate strip chart. Each chart division was equivalent to about 2 pphm. Since the signal results from several transfer functions, accuracy can be obtained only on the basis of dynamic calibrations. It is not unusual to find the vendor's calibrations to be low by 25% based on the colorimetric ozone procedure (15,16).

Colorimetric Oxidant Analyzers

Sample air is drawn at a metered rate into a contact column where the air is scrubbed with a metered flow of potassium iodide buffered at pH 6.8. The reaction of oxidants with the KI solution produces triiodide ion (I_3^-) which then passes to a colorimeter cell where the absorbance of the I_3^- is measured at 354 nm (7). A diagram of a colorimetric oxidant analyzer is shown in Figure 2. The analyzer's detector signal was recorded on a 10-inch wide semi-log strip chart in terms of absorbance. The chart graduations, therefore, are expanded at low signal levels and compressed at maximum signal levels in contrast to the equi-interval divisions of the recorder charts from the amperometric analyzer.

As shown in Table I, two types of air-liquid scrubbers were used: 1) an inserted helix and 2) a helical coil. The helix insert type column, used with the 10% KI, consisted of a glass tube 5 to 6 mm ID either 23 or 50 cm long in which a glass helix is placed axially and in contact with the inside surface of the column wall. The liquid passes down the column in a thin film while the sample air flows countercurrent up the tube. The analyzer equipped with the 50 cm column was used in Phase I and the 23 cm column was used in Phase II. While the shorter column is about half the length of the longer, the nominal air and reagent flow rates were proportionately less thus maintaining the same ratio. Tests conducted showed no significant differences between these two columns in scrubbing efficiencies for O_3 and NO_2.

The helical coil column, used with the 20% KI, was fabricated from 5 or 6 mm ID glass tubing which had been coiled 2-1/2 to 3 times around a cylindrical form about 10 cm in diameter. It was mounted with the axis vertical and the air-liquid flow direction was concurrent up. Some users have modified the flow to concurrent down.

Colorimetric NO_2 Analyzer

Analysis for NO_2 was conducted by scrubbing sample air with an absorbing solution containing 1.5% 2-aminobenzenedisulfonic acid (ABDS), 0.001% N(2-napthyl)ethylenediamine dihydrochloride (NEDA), 15% ethylene glycol and 0.05% sodium benzoate (8). NO_2 reacts with NEDA to form a nitroso

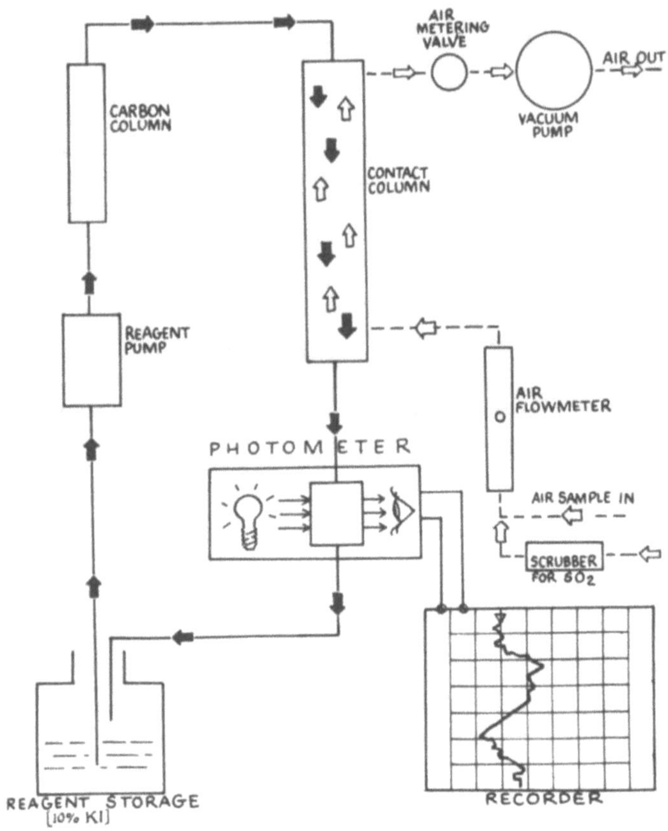

Figure 2. OXIDANT ANALYZER - COLORIMETRIC

compound which then reacts with the ABDS to form a pink azo-dye. The absorbance of the dye was recorded continuously at 550 nm on a 2½-inch wide rectangular coordinate strip chart. A schematic diagram of an NO_2 analyzer is shown in Figure 3.

NO_2 INTERFERENCE EQUIVALENT

Nominal concentrations of ozone (O_3) between 0.0 to 60 pphm were generated by irradiating filtered air with ultraviolet light. The air was filtered through the same charcoal-soda lime cartridge (13) as described in the calibration procedure (10). The oxidant analyzer sampled ozone only until the recorder indicated a constant response. The actual O_3 concentration was then determined manually (9).

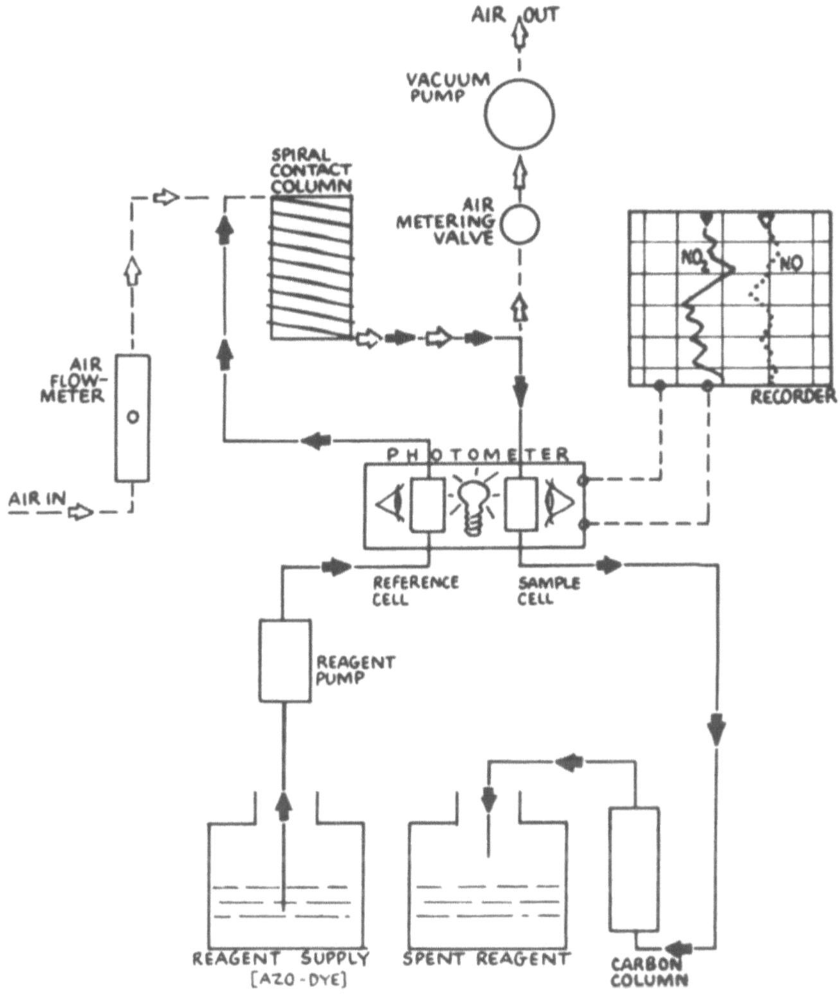

Figure 3. NITROGEN DIOXIDE ANALYZER - COLORIMETRIC

Concentrations of NO_2 in the range from 0.0 to 60 pphm were then introduced into the O_3 stream and the new instrument reading noted. The acutal NO_2 concentrations were also determined manually (11). The oxidant readings were adjusted for the decrease in oxidant concentration due to the additional volume contributed by the NO_2. We confirmed under the conditions of this study, O_3 did not interfere in the determination of NO_2 and the time for reaction between NO_2 and O_3 was insufficient to cause any significant artifacts. The system used to generate and vérify the O_3 and NO_2 streams is shown in Figure 4.

The increase (y) in oxidant reading due to NO_2 was determined by subtracting the ozone concentration, $[O_3]$, from the oxidant concentration $[O_x]$, after NO_2 was added:

$$y = [O_x] - [O_3]$$

A regression line was calculated for the change in (y) on the NO_2 concentration (designated x). The slope of this line is equivalent to the fraction of the NO_2 appearing as oxidant. This mean interference equivalent (*f*) was estimated from the ratio $\Sigma y/\Sigma x$ or:

$$f = \frac{\Sigma \left\{ [O_x] - [O_3] \right\}}{\Sigma \, [NO_2]}$$

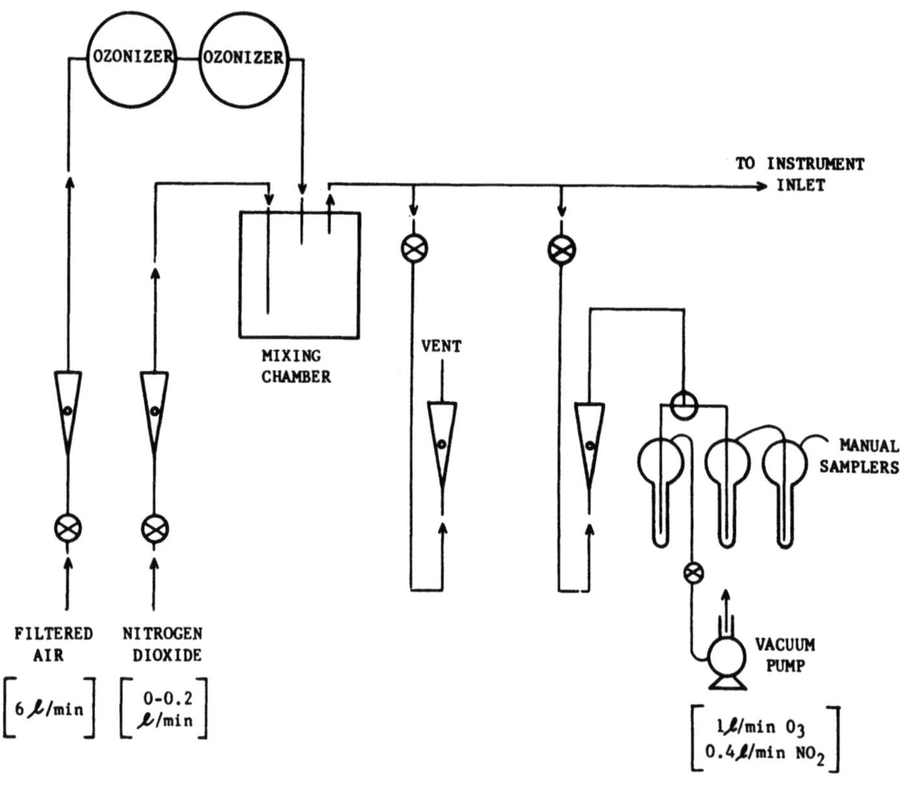

Figure 4. FLOW DIAGRAM - OZONE GENERATION AND GAS DILUTION SYSTEM

The data presented in Tables II, III, and IV for the amperometric and colorimetric analyzers show NO_2 in the sampled air increases the iodometric oxidant response and that this increase is more pronounced with higher KI concentrations. For the amperometric analyzer the response varied from 0.025 [NO_2] at the highest O_3 level to 0.081 [NO_2] when O_3 was absent. Analogous variations from 0.179 to 0.223 occurred with the colorimetric 10% KI and from 0.189 to 0.342 with 20% KI.

Table II

AMPEROMETRIC 2% KI – RESPONSE TO NO_2, pphm

OZONE LEVEL [O_3]	NO_2 INPUT	ANALYZER RESPONSE [O_x]	INTERFERENCE EQUIVALENTS: Each Test	INTERFERENCE EQUIVALENTS: Each O_3 Level†	INTERFERENCE EQUIVALENTS: Overall†
0.0	4.0	0.0	0.0	0.081	0.0627
	11.0	0.4	0.036		
	29.5	2.5	0.085		
	54.5	5.1	0.094		
13.1	5.0	13.1	0.0	0.076	
	9.0	14.0	0.100		
	22.0	14.7	0.073		
	32.0	15.4	0.072		
	52.0	17.2	0.079		
	59.5	17.8	0.079		
25.3	7.5	25.8	0.067	0.053	
	11.5	25.6	0.026		
	32.0	27.2	0.059		
	55.0	28.2	0.053		
60.9	8.0	61.1	0.025	0.025	
	10.5	61.1	0.019		
	33.5	61.6	0.021		
	62.5	62.7	0.029		

† Average interference equivalents, *f*, determined by least squares.

To establish whether such interactions were of practical significance, an overall least squares fit interference value *f* was determined for each analyzer. The *f* values calculated are 0.0627 for the amperometric unit, 0.2161 for the 10% KI colorimetric and 0.317 for the 20% KI colorimetric units. As expected from the data, the greatest deviation from these best-fit lines occurred at the highest O_3 levels.

Table III

COLORIMETRIC 10% KI* – RESPONSE TO NO_2, pphm

OZONE LEVEL $[O_3]$	NO_2 INPUT	ANALYZER RESPONSE $[O_x]$	INTERFERENCE EQUIVALENTS Each Test	Each O_3 Level†	Overall†
0.0	12.7	2.6	0.205		
	23.5	4.9	0.209		
	30.8	6.6	0.214	0.220	
	38.0	8.3	0.218		
	57.9	13.5	0.233		
9.7	14.5	12.2	0.172		
	24.4	14.8	0.209	0.223	
	39.8	19.1	0.236		
	59.7	23.6	0.233		0.2161
29.0	14.5	31.7	0.186		
	25.3	34.5	0.217	0.223	
	41.6	38.1	0.219		
	58.8	43.0	0.238		
54.7	14.5	56.9	0.152		
	27.2	59.1	0.162	0.179	
	45.3	62.2	0.166		
	62.4	67.4	0.204		

* 20-inch insert helix column.

† Average interference equivalents, *f*, determined by least squares.

Table IV

COLORIMETRIC 20% KI* – RESPONSE TO NO_2, pphm

OZONE LEVEL $[O_3]$	NO_2 INPUT	ANALYZER RESPONSE $[O_x]$	INTERFERENCE EQUIVALENTS Each Test	INTERFERENCE EQUIVALENTS Each O_3 Level†	INTERFERENCE EQUIVALENTS Overall†
0.0	4.0	0.8	0.200	0.332	0.3170
	11.0	3.3	0.300		
	29.5	9.2	0.312		
	54.5	19.6	0.360		
13.3	5.0	13.3	0.0	0.342	
	9.0	15.2	0.211		
	22.0	20.0	0.305		
	32.0	24.0	0.334		
	52.0	33.2	0.383		
	59.5	35.5	0.373		
25.1	7.5	26.7	0.213	0.299	
	11.5	27.6	0.217		
	32.0	34.1	0.218		
	55.0	43.7	0.338		
61.3	8.0	61.5	0.025	0.189	
	10.5	61.6	0.028		
	34.5	67.1	0.618		
	62.5	76.8	0.218		

* Helical coil column.

† Average interference equivalents *f*, determined by least squares.

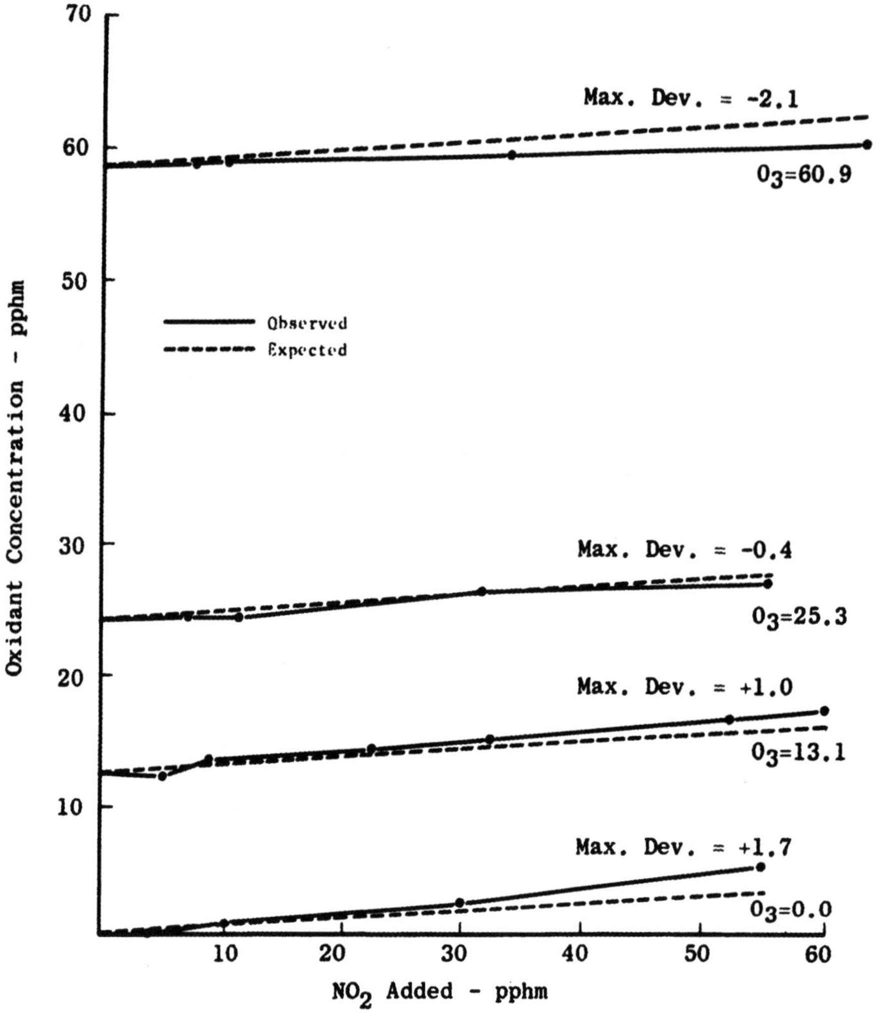

Figure 5. AMPEROMETRIC - 2% KI - OXIDANT CONCENTRATION OBSERVED AND EXPECTED. BASED ON 0.0627 NO_2 INTERFERENCE.

This is shown in Figures 5, 6 and 7 for the three analyzers. At the highest O_3 levels the average calculated interference estimates greater interference from NO_2 than that actually occurring. This leads to an over correction of the oxidant reading by approximately 4% of the actual O_3 value with the amperometric and colorimetric 10% KI analyzer (Figure 6). Statistically, the differences between the observed and expected values were not significant.

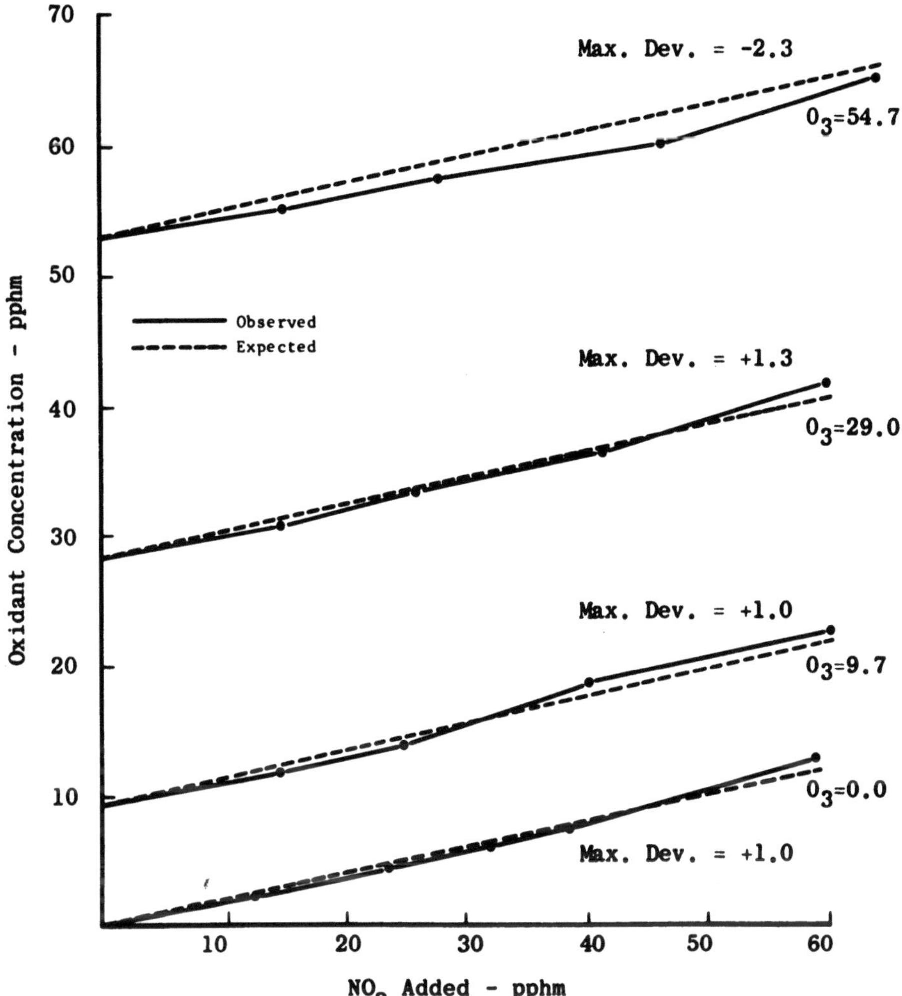

Figure 6. COLORIMETRIC - 10% KI - OXIDANT CONCENTRATION OBSERVED AND EXPECTED. BASED ON 0.2161 NO_2 INTERFERENCE

The differences, however, were much more pronounced with the 20% KI procedure (Figure 7). The over correction in this case was as high as 13% and these differences are statistically significant by the Wilcoxon Signed Rank Test ($p < .05$). Therefore, no overall correction can be applied. When the 20% KI reagent is utilized a different factor for each ozone concentration interval should be applied. This reagent has not been used by most California air monitoring agencies since early 1968 and such determinations would only be necessary for comparisons including pre-1968 data.

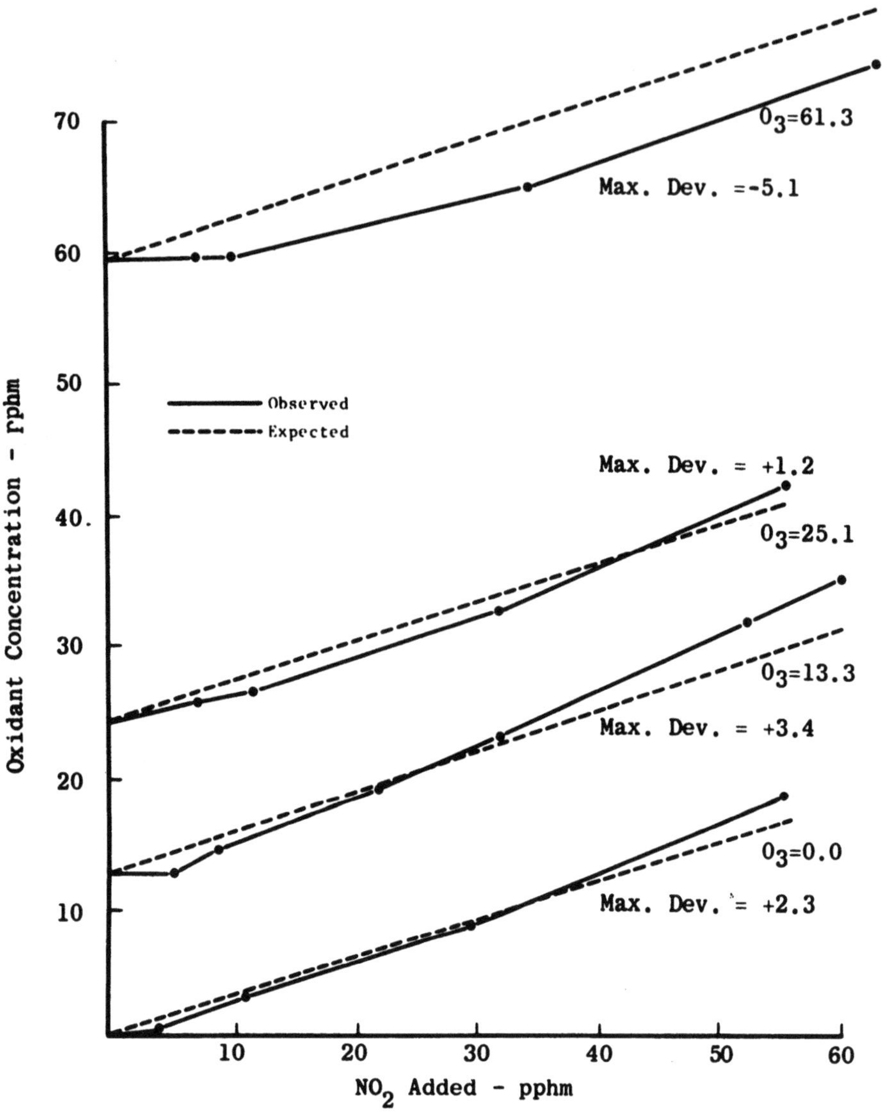

Figure 7. COLORIMETRIC - 20% KI - OXIDANT CONCENTRATION OBSERVED AND EXPECTED. BASED ON 0.3170 NO_2 INTERFERENCE

To what extent differences in reagent formulations and contactor design between 10% and 20% KI instruments contributed to the observed differences in the interference equivalents was not determined in this study. Work done with similar contactors indicate the KI concentrations change the NO_2 interference equivalent (1,17). However, we have shown in unpublished work that contactor deign is also of importance. For instance, using the 10% KI reagent, an air to liquid flow ratio of 0.4 l/ml and a 10 cm packed column, concurrent down, yielded an NO_2 interference equivalent of about 0.36. Since the solubility of NO_2 in KI solutions is much less than O_3, it should be possible to reduce the iodometric response to NO_2 by optimizing reagent formulation and contactor design.

VERIFICATION UNDER MONITORING CONDITIONS

To collect air monitoring data for the verification phase of this study, the amperometric and the colorimetric 10% KI analyzers were operated side-by-side for a period of 75 days during the period August through October, 1969 in Pasadena, California. The instruments were located in an air-conditioned laboratory and connected to a common sampling manifold. The ambient air was drawn from an elevation of about 18.3 m and was transported through a vertical polyvinyl chloride pipe 6.3 cm ID by 20.4 m long to the glass distribution manifold. Losses of oxidant and NO_2 were checked by collecting manual samples simultaneously near the inlet of the sampling pipe and from the manifold. The losses were less than 10%.

The operation of all analyzers was checked at intervals throughout the working day. Filtered air was introduced daily at the start of each day (between 0800 to 0900 hours) and instrument zeros were reset when necessary. During the final few minutes of the zeroing period, the response of the colorimetric detectors was checked with optical filters and any needed corrections were noted. Once a week all air flow rates were validated with a wet test or soap bubble meter; the liquid flow rates were validated by measuring the volume of solution delivered during a fixed time interval. Calibrations were checked dynamically at approximately two-week intervals. Reagents were replenished as required and replaced every two weeks.

The recorder charts were removed every 24 hours. The readings were averaged over each one hour interval for each hour of the day and the peak reading for the day was recorded. These values were then corrected for changes in the instrument zero and shifts in calibration. Based on the concurrent NO_2 concentrations, the amperometric and colorimetric oxidant values were corrected for the NO_2 interference by subtracting the contribution of the NO_2 to the oxidant reading as follows:

$$[O_3] = [O_x] - \bar{f}[NO_2]$$

where: $[O_3]$ = oxidant concentration corrected for NO_2 interference

$[O_x]$ = analyzer reading

$[NO_2]$ = concurrent NO_2 concentration

$\bar{f}$ = interference equivalent (0.0627 for amperometric; 0.2161 for colorimetric as determined in Phase I)

All readings during periods when one analyzer was not operating were then eliminated from consideration. The missing data resulted during periods when one or more of the analyzers was not operating due to zeroing, calibration or malfunction.

A total of 1,797 hours of actual monitoring data was available. During this period the analyzers operated concurrently for 1,551 hours. Of these, valid amperometric and colorimetric measurements corrected for NO_2 interference yielded data for 1,170 hours, which were used to compare the analyzers.

The distribution of the corrected hourly averages at various oxidant concentration levels is shown for each analyzer in Figure 8 in the form of a histogram. The colorimetric readings tended to be higher than the amperometric. The maximum hourly average (corrected for NO_2 interference) recorded by the colorimetric unit was 53 pphm; the amperometric 47 pphm. The mean hourly average during the study period for the colorimetric was 5.4 pphm, somewhat higher than the amperometric at 4.7 pphm. The distributions are significantly different by the Kolmogorov-Smirnov two-sample test ($p < .001$).

The corrected hourly averages are compared point by point in the form of a scatter diagram in Figure 9. The instruments, when corrected for NO_2 interference were highly correlated ($r = 0.96$). The plot also indicates the overall colorimetric readings were consistently higher than the amperometric readings by about 6% even after correction for NO_2. The relationship is described by the following equation:

$$[O_3]_A = 0.942[O_3]_C - 0.375 \text{ pphm}$$

where: $[O_3]_A$ = amperometric reading corrected for NO_2

$[O_3]_C$ = colorimetric reading corrected for NO_2

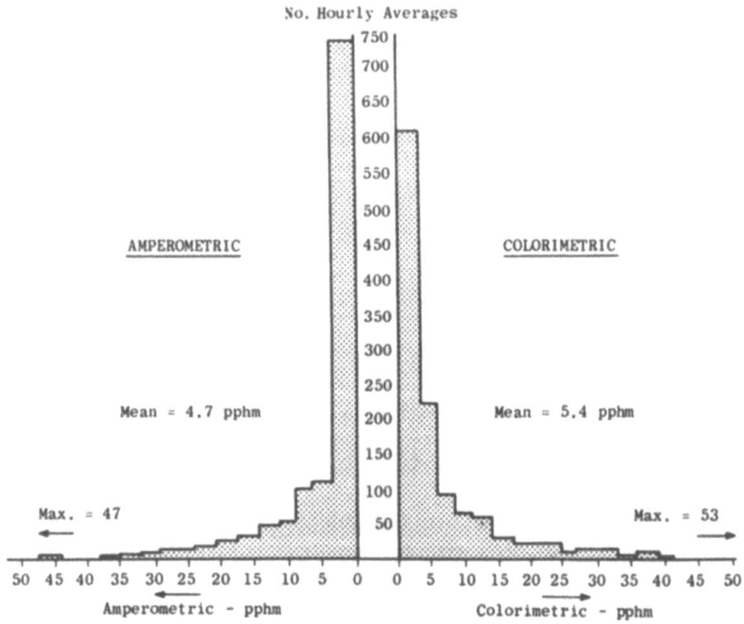

Figure 8. NUMBER OF HOURLY AVERAGES AT EACH OXIDANT CONCENTRATION.

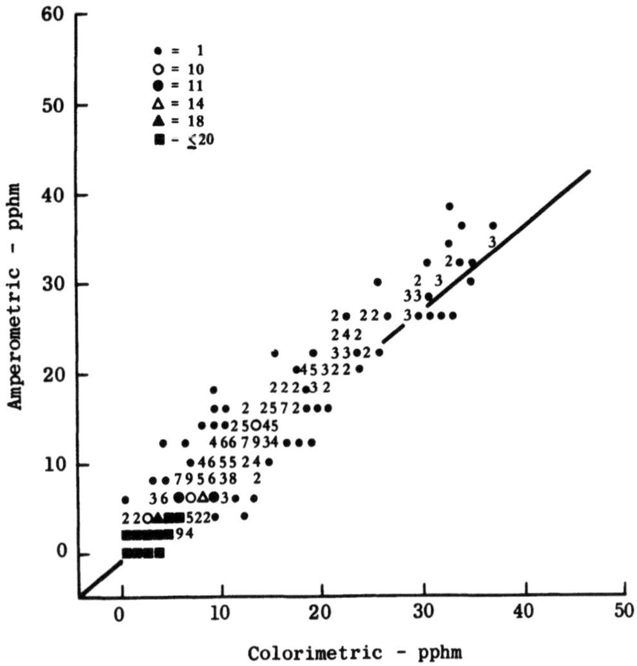

Figure 9. AMPEROMETRIC-COLORIMETRIC COMPARISON OXIDANT HOURLY AVERAGES CORRECTED FOR NO_2 INTERFERENCE

Because individual values occasionally differed from each other by as much as 100%, correlation of diurnal patterns produced by each analyzer might provide better estimates of comparability than the overall correlation.

The mean hourly average for these analyzers for each hour of the day is shown for uncorrected data in Figure 10. The same data, corrected for NO_2, are shown in Figure 11. Superimposed on both plots are the non-parametric correlations (Spearman's ρ) as well as the mean concurrent hourly NO_2 concentrations for each hour of the day.

Figure 10 shows the magnitude of the disparity (about 1.6 pphm) between uncorrected data collected by the amperometric and colorimetric analyzers due to the unequal contribution of NO_2 to each in the oxidant-KI reaction. Figure 11, however, demonstrates the extent to which the disparity due to NO_2 interference can be reduced by application of interference equivalents established by the laboratory investigation in Phase I. In this case, the mean uncorrected difference of about 1.6 pphm was reduced by half to about 0.7 pphm. The graphs, however, show the colorimetric readings were consistently higher even after correction for the NO_2. Although the difference after correction between the two instruments was most pronounced during the midday peak similar differences exist during the low night and early morning hours.

The two graphs also demonstrate the high overall correlation found between the two instruments shown previously in Figure 9. The correlation was particularly high (ρ = 0.88 to 0.95) during the midday peak even though the absolute difference between the two instruments was maximal during this period. The correlation, however, was very poor at the inflection points in the diurnal pattern where the oxidant level begins to rise (0600 to 0700) and again as the oxidant concentration levels off in the evening (2000 to 2200).

DISCUSSION

While the measurements made by the two analyzers were highly correlated, the correlation was not consistent throughout the oxidant concentration range (ρ = 0.0 to 0.97). The colorimetric unit read between 0.3 to 1.0 pphm higher even after correction for NO_2. While the differences between the readings are considered to be within the precision of the analytical technique (± 1 pphm), these observed differences were nevertheless statistically significant.

At the lower concentration levels one contributing factor to the consistently higher colorimetric readings may be a lack of precision in reading the amperometric recorder chart. The wider colorimetric analyzer chart is expanded at low oxidant levels and is much easier to read accurately. Thus, variations of 1 or 2 pphm which register as slight wiggles on the amperometric trace, register as definite excursions with the colorimetric trace. This could result in under reading the

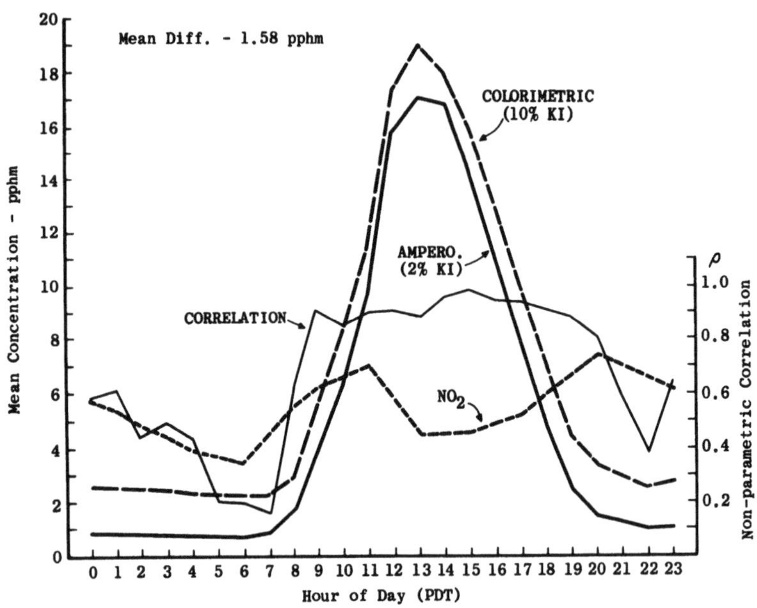

Figure 10. MEAN HOURLY AVERAGES FOR EACH HOUR OF DAY (Oxidant not corrected for NO_2 interference)

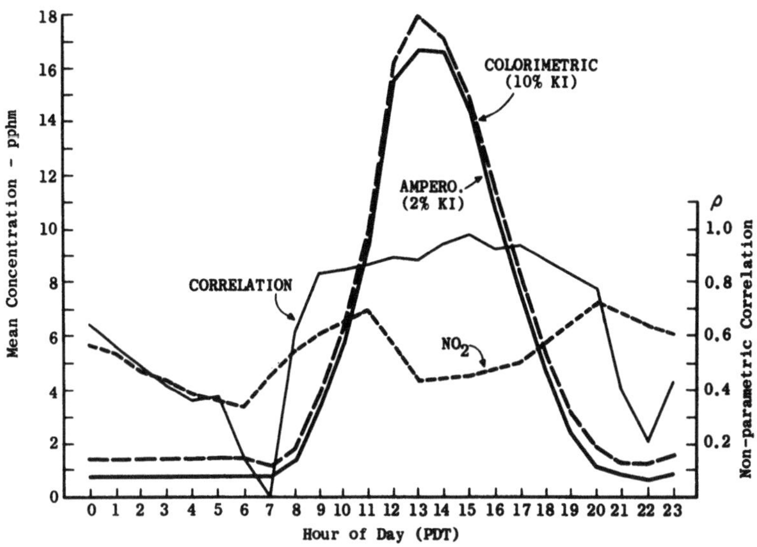

Figure 11. MEAN HOURLY AVERAGES FOR EACH HOUR OF DAY (Oxidant corrected for NO_2 interference) .

amperometric charts at low levels and contribute to the amperometric-colorimetric differences there.

However, this does not explain the higher colorimetric values at the peak. Since the interference of oxidants such as NO_2 is greater for 10% KI compared to 2% KI, it is possible other oxidants (e.g., hydrogen peroxide) will interfere in the same fashion. Up to 20 pphm hydrogen peroxide have been observed in Riverside near Pasadena (18).

Also, the correlation was poor during periods of rapid oxidant concentration change. One possible explanation for this is the difference in analyzer response times. The amperometric unit is able to register 90% of an abrupt change in ozone concentration within one to two minutes; the colorimetric unit requires 10 to 15 minutes. An example of the slower response is seen between 0600 to 0700 where the amperometric readings are beginnning to rise from 0.7 to 0.8 pphm while the colorimetric signal is still decreasing from 1.5 to 1.3 pphm. The data were not adjusted for the differences in response time.

Analyzer precision tends to be poorest at concentrations near the lower limits of detectability. Because the colorimetric response is greater than the amperometric at low oxidant levels, a small decrease in concentration would be registered by the colorimetric unit and not by the amperometric unit. Such variations at low levels would be proportionately large with respect to the overall reading and would contribute adversely to the correlation between the two instruments.

Eliminating from consideration all hourly averages when one or the other analyzer was not operating, we have 1,551 hours of uncorrected data when both analyzers were operating. For 113 hours (an additional 7%) the colorimetric data are missing. Most of these (65%) occurred during daylight hours and in particular during the 0800 to 0900 period. The analyzers were usually zeroed at the beginning of the work day. The faster response of the amperometric unit minimized "off-stream" time, whereas the slower colorimetric unit was "off-stream" for nearly an hour.

The uncorrected amperometric readings during the period of missing colorimetric data averaged 7.9 pphm as opposed to the overall uncorrected amperometric average of 5.2 pphm; this is significantly higher ($0.001 < p < 0.005$) by the Kolmogorov-Smirnov two-sample test. The amperometric analyzer was inoperable for 76 hours, but the colorimetric readings during this period were not significantly different from the overall colorimetric data.

Thus a bias may exist which would increase the magnitude of the amperometric-colorimetric difference. The overall effect of having to eliminate the 113 hours from our evaluation would be a bias toward under-representing the high levels recorded by the colorimetric unit and the true difference between these two analyzers might be greater than we have observed.

Although in Southern California oxidant concentration may reach 100 pphm, NO_2 values rarely exceed one-third of this. Since the NO_2 and O_3 levels encountered during Phase II did not exceed the range investigated in the determination of the NO_2 interference equivalent, application of the average interference to the ambient air data is considered to be realistic.

Although the differences between the amperometric and colorimetric readings are significant, the absolute differences are small after correction for NO_2 interference ($\leqslant$ 1.0 pphm). Random differences of this magnitude might exist between monitoring data collected by two identical analyzers operating side-by-side. However, the consistent pattern of higher colorimetric readings we have observed would seem unlikely. This situation indicates an air monitoring station may over- or under-represent the true situation even after correction for NO_2 interference depending on the type of analyzer and reagent formulation. The differences after correction between these two instruments would not significantly alter the frequency of reporting violations of air quality standards. Nor would the data create significant differences in assessing the total exposure of an individual to high oxidant levels.

SUMMARY AND CONCLUSIONS

The magnitude of the increase in oxidant reading due to NO_2 when using iodometric analyzers is dependent on reagent formulation, analyzer contact column design, levels of NO_2 and O_3 and perhaps detection principle. The response is greater with increasing iodide concentration and contact column configurations which improve gas to liquid transfer. It is possible to control the iodometric response to NO_2 by optimizing reagent formulation and contactor design.

Even after correction for interferences the data may not correspond exactly due to differences in instrument performance, atmospheric composition, maintenance and calibration practices and frequencies, analyzer environment, and data acquisition techniques.

Interference equivalents have been established for three types of widely used oxidant analyzers. Any change in reagent or contactor design, even those considered trivial, require confirmation by laboratory and field comparisons.

The opportunities for other pollutants to interfere in the iodometric procedure for oxidant are many. In locations where significant SO_2 concentrations do not occur, the current wet-chemical KI systems for monitoring oxidants and ozone in air are adquate to provide information concerning air quality and to record violations of air quality standards provided the readings are corrected for the interference from NO_2. Ideally, it would be best always to correct for both NO_2 and SO_2.

Acknowledgements

This work was supported in part by U.S.E.P.A., APCO, Contract CPA 70-24, Mr. Thomas Stanley, Project Officer; and by the California Air Resources Board, Mr. John Maga, Executive Secretary.

REFERENCES

1. Cherniack I, Bryan RJ: A comparison study of various types of ozone and oxidant detectors which are used for atmospheric sampling. Proc 57th Annual Meeting of the Air Pollution Control Association (June, 1964)

2. Mueller PK, Terraglio PK, Tokiwa Y: Chemical interferences in continuous air analysis. Proc 7th Conf Methods in Air Pollution Studies, Berkeley, Calif. (January, 1965)

3. Siu W, Ahlstrom MR, Feldstein M: Comparison of the coulometric and colorimetric oxidant analyzer data. Proc 10th Conf Methods in Air Pollution and Industrial Hygiene Studies, Berkeley, Calif. (February, 1969)

4. Unpublished data. Air and Industrial Hygiene Laboratory, California Department of Public Health, Berkeley, Calif.

5. Hodgeson JA, Martin BE, Baumgardiner RE: Comparison of chemiluminescent methods of measurement of atmospheric ozone. Vol. V, Plenum Press, 1971

6. Mast Development Company: Ozone Meter Model 725-6, Davenport, Iowa

7. Beckman Instruments, Inc: Air Quality Acralyzer Model 76 and 77, Fullerton, Calif.

8. Atlas Electric Devices Company: NO_2 Analyzer Model 1302, Chicago, Illinois

9. Air and Industrial Hygiene Laboratory: Total Oxidant Content of the Atmosphere -- 2% Potassium Iodide, Recommended Method No. 2-A. California State Department of Public Health, Berkeley, 1969

10. Air and Industrial Hygiene Laboratory: Calibration and Standardization of Continuous Photometric Analyzers of Atmospheric Oxidants, Recommended Method No. 29. California State Department of Public Health, Berkeley, 1970

11. Air and Industrial Hygiene Laboratory: Nitrogen Dioxide and Nitric Oxide Content of the Atmosphere, Recommended Method No. 3. California State Department of Public Health, Berkeley, 1961

12. Air and Industrial Hygiene Laboratory: Calibration and Standardization of Atmospheric Nitrogen Dioxide and Nitric Oxide Continuous Analyzers, Recommended Method No. 34. California State Department of Public Health, Berkeley, 1970

13. Mine Safety Appliance: Filter GMC CR-10340 (for organic vapors and acid gases), Pittsburg, Pa.

14. Intersociety Committee Manual of Methods of Air Sampling and Analysis: Tentative method for continuous monitoring of atmospheric oxidants with amperometric instruments. Health Lab Sci 7:1, Jan 1970

15. Gudiksen PH, Hildebrand PW, Kelly JJ: Comparison of an electrochemical and a colorimetric determination of ozone. J Geophys Res 7:22, Nov 66

16. Wartburg AF, Brewer AW, Lodge JP: Evaluation of a coulometric oxidant sensor. Int J Air & Wat Poll 8:21, 1964

17. Littman FE, Benoliel RW: Continuous oxidant recorder. Anal Chem 25:1480, 1953

18. Bufalini JJ: Private communication, March 1971

THE DETERMINATION OF TRACE METALS IN AIR

Philip W. West

Coates Chemical Laboratories

Louisiana State University, Baton Rouge, La. 70803

Environmental pollution has become a matter of great concern to all sections of this country, affecting rich and poor alike. The nature, extent, and the ultimate control of pollution requires reliable analytical methods capable of accurately delineating the problem in all of its aspects. All too often "black boxes" are used to measure pollutants. The instruments used may provide valid information but more often than not they measure some general, rather than specific, property of the species to be determined. Obviously such devices lack specificity and therefore may provide misleading and erroneous data. Finally, there is an unfortunate tendency to introduce exotic methods and expensive instruments which are beyond the capability and budgets of many concerned organizations.

The analysis of air to determine the concentrations of significant trace elements is of special interest. Arsenic, chromium, and nickel are known to be carcinogenic to man; and beryllium, cadmium, cobalt, lead, and selenium have been indicted as possible carcinogens to man because of their demonstrated carcinogenicity to experimental animals (1). All of these elements may also be considered toxic, at least in high concentrations. To the above list of toxins can be added such elements as antimony, barium, manganese, mercury, silver, tin, and vanadium (2).

Adding even more significance to the trace elements is a knowledge that concentration, form, mode of introduction, and the presence or absence of co-pollutants may alter the significance of a given element. Selenium (3), for example, causes "blind staggers," and fatalities due to the so-called "alkali disease" when it is ingested in relatively high concentrations (10-50 ppm) by foraging animals. Similar concentrations have been found to cause cancer of the liver when fed to experimental animals. On the other hand, the lack of dietary selenium may cause muscular dystrophy and "white muscle disease." Manganese, which is considered an essential element when introduced by ingestion, is considered to be a causative agent of Parkinson's disease when inhaled. Iron is essential in human nutrition, yet it has become a suspected carcinogen when introduced as iron dextran and similar preparations (1).

From the foregoing summary it is clear that there is an urgent need for sensitive, reliable analytical methods that are applicable to the study of environmental pollution. The methods required can be classified into three categories. The first group are the reference methods that provide reliable, critical data regardless of the complexity and cost. The second group is made up of field methods and simple procedures suitable for use by less highly trained personnel in laboratories where budget limitations preclude the use of highly sophisticated equipment and procedures. A third category consists of monitoring methods which involve automatic instruments for continuous measurements of pollutants, preferably with telemetering capabilities. The latter category is not included in this discussion.

Two complimentary analytical approaches are proposed which provide both reference methods and field or general application procedures. Critical studies of most trace metals are best carried out by means of atomic absorption spectroscopy, which can provide reference values for base-line monitoring, studies involved in litigation, etc. The recommended alternate approach is the use of the ring oven technique, which gives highly reliable, although less accurate, data but has the advantage of being remarkably simple, cheap, and rapid. It is well suited for field studies and for use in laboratories where sophisticated equipment and highly specialized personnel are unavailable.

REFERENCE METHODS OF ANALYSIS

Atomic absorption spectroscopy has been accepted as one of the most ideal analytical techniques known. Its inherent high degree of specificity together with its simplicity and relatively low cost have established it as the method of choice for the determination of most metals and even some of the metalloids. Although advances are still being made in the design of equipment and in establishing optimum conditions for spectroscopic measurements, it is obvious that atomic absorption spectroscopy can be applied with confidence for most analyses of properly presented environmental samples. The important question of the sample, however, is still not completely resolved. To picture the complexity of the problem of samples, it is pointed out that most airborne particulates are collected by means of high-volume samplers using glass filter sheets as the collection medium. Unfortunately, the glass filters are excessively contaminated with a number of trace metals and thus introduce high and unfortunately, variable blanks. When an acid leach is used to remove the particulate species from such filters, variable amounts of the filter are dissolved, even when careful control of acidity, temperature and time is maintained. Monkman, in the Environmental Health Center of Canada, has recognized this problem and has gone to the complete dissolution of the filters and particulates using a hydrofluoric acid treatment (4). A high, but reasonably constant blank is encountered with this approach. An alternative approach is suggested which offers the advantages of low blanks, convenience and rapidity (5). Instead of glass filters, acid washed filter papers are recommended for sample collection. The collected samples are then divided and a known fraction taken for analysis. If, for example, 4 inch circles of filter paper are used for the sampling, they may be quartered and opposite sectors placed on the bottom of a 100 ml beaker. Two milliliters of 15% ammonium acetate are then added followed by twenty milliliters of an ethyl propionate solution of mixed ligands (0.1 g diphenylthiocarbazone; 0.75 g 8-quinolinol; and 20 ml acetyl acetone dissolved and made up to 100 ml with ethyl propionate). The mixture is then equilibrated employing occasional shaking, whereby the trace metals are dissolved and extracted into the ethyl propionate as mixed chelates.

The samples collected and processed as described above are in suitable form for atomic absorption spectroscopy. The metal chelates are separated from anions, such as phosphates, that occasionally interfere with direct spectroscopic measurements. Also, atomization of the ethyl propionate solutions results in a 5-10 fold enhancement of sensitivity over sensitivities obtained with aqueous systems.

Sampling and Analysis

Sample: Collect 25 m^3 (or other suitable volume) of air using a high-volume sampler and acid-washed filter paper as the sampling medium. Divide the filter with the particulate sample, taking a suitable fraction and treating it first with ammonium acetate and then with mixed ligand solution as described above.

Analysis (5): Aspirate directly the ethyl propionate solution of the mixed metal chelates using a HETCO burner with triflame burner head (Jarrel Ash Co.), or equivalent, and a suitable atomic absorption spectrophotometer, such as the Perkin-Elmer model 303. Table I provides the necessary operational details.

TABLE I

OPTIMUM CONDITIONS FOR ATOMIC ABSORPTION MEASUREMENTS

Metal	Wavelength (Å)	Spectral slit width (Å)	Lamp-current (mA)	Acetylene flow*
Ag^{+}	3281	2.0	12	2.2
Cd^{2+}	2288	6.5	8	2.5
Co^{2+}	2407	2.0	20	2.0
Cu^{2+}	3247.5	2.0	10	2.0
Fe^{3+}	2483.3	2.0	12	2.5
Ni^{2+}	2320	2.0	14	2.0
Pb^{2+}	2170	6.5	30	2.0
Zn^{2+}	2138	2.0	15	2.0
Be^{2+}	2348.6	2.0	20	21†

* Arbitrary scale units on Hokes flowmeter model 993 with an air flow of 10 at 35 p.s.i.
† With a nitrous oxide flow of 10 at 35 p.s.i.

Working curve: Prepare series of standard working curves by taking appropriate volumes of standard metal solutions to provide reference points for the following ranges (μg metal/ml): Ag, 0.2-2.5; Cd, 0.1-1.0; Co, 0.4-2.0; Cu, 0.2-2.5; Fe, 0.4-4.0; Ni, 0.4-3.0; Pb, 0.5-6.0; Zn, 0.1-1.0; Be,0.2-1.0. To each standard increment add 2 ml of ammonium acetate and 20 ml of mixed ligand solution. Extract and atomize as described under "analysis." Plot the respective absorbances to provide the working curves.

Calculations: Report results in terms of micrograms of metal per cubic meter of sample. Sample volume must be corrected for fraction used for extraction and analysis. The micrograms of metal found are obtained directly from the working curve.

Standard metal solutions: Standard stock solutions of metals are best prepared by accurately weighing pure metal, dissolving in a minimum of concentrated nitric acid, and diluting to known volume with distilled water. Most metals are readily obtainable four nines pure.

Note: The procedure described above is applicable to the study of trace metals in usual samples of airborne particulates. Where metal sulfides are to be included, the collected samples should first be exposed 10 minutes to bromine vapors to convert the sulfides to sulfates. Excess bromine should be removed by exposing the filters to a gentle flow of warm air before final treatment with ammonium acetate and the mixed ligand solution.

GENERAL UTILITY METHODS

The ring oven technique introduced by Weisz (7) provides a simple, elegant method for the concentration of small amounts of material to permit increased sensitivity and quantification of spot reactions. In addition, by various schematic procedures, a variety of types of separations may be carried out on a single spot of sample unknown prior to final spot test identification and/or determination. The discussion which follows will present manipulative details for using the ring oven in several types of procedures. The method is ideally suited for the analysis of airborne particulates because it adapts directly to the separation, concentration, and determination of trace particulate species as collected on filter tapes with sequential samplers. The individual procedures that have been developed and included here are highly selective and reliable. Errors in the quantitative measurements seldom exceed 20%, thus making the method comparable with other trace analysis techniques such as emission spectroscopy and polarography.

The general design of the ring oven is shown in Figure 1. A commercial model is now available from Arthur H. Thomas & Co., Philadelphia, under the trade name "Trace Oven." The Trace Oven has a built in rheostat for adjusting the temperature, and it is supplied with a choice of auxiliary rings which contributes to the flexibility of its use.

The capillary pipet illustrated is the "solvent pipet" supplied with the commercially available equipment. To use conventional micropipets the guide tube may be replaced by one machined to accept the normal micropipet. Teflon presents many advantages as a material for this modification. The guide tube must be so positioned that the tip of the pipet strikes the center of the annular space in the heating block. The retainer ring serves simply to hold the filter paper in place.

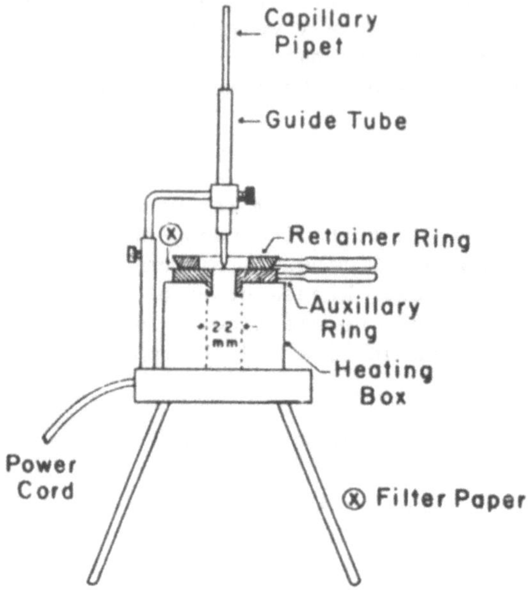

Fig. 1. Ring oven.

The operating principle of the ring oven is based on the addition of solutions and solvents to the center of filter paper. The liquids then diffuse through the pores of the paper to the inner edge of the heating block where the solvent is evaporated. Successive additions of solvent, via the pipet, wash the solute to the ring zone (inner edge of the heating block), effecting a fifty to one hundred-fold concentration of the solute.

The auxiliary ring permits a preliminary concentration of solute at a smaller ring zone, from which, by appropriate choice of solvents, selected components of the solute mixture may be extracted and washed to the normal ring zone after removal of the auxillary ring. A second, larger, auxillary plate may be used to permit further separations by washing unwanted material outward from the normal ring zone. Once the ring is formed and the paper removed from the ring oven, additional chemical steps may be taken. The filter paper may be sprayed with and/or dipped in additional reagents to achieve desired reactions permitted by the chemistry of the system, such as developing a color or washing out excess reagents or interfering species.

For qualitative purposes, the steps discussed above should suffice. For the quantitative estimation of traces, the above operations may again be followed and quantification achieved by comparing the test ring with standard rings prepared in the same manner. The sequence of operations is summarized in Figure 2.

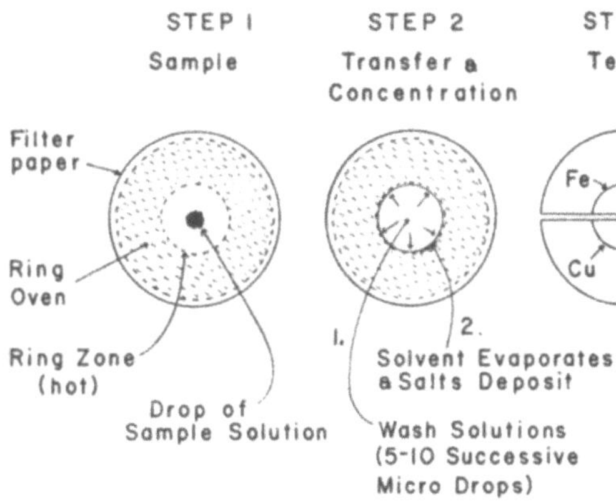

Fig. 2.

Sampling and Analysis

The ring oven technique provides a very convenient means for evaluating the nature and amount of airborne particulates. Preferably, samples should be collected by means of a sequential tape sampler (eg., Unico, 80 TS or the Gelman Air Sampler) and the filter tapes used as the reaction medium for all subsequent processing of the collected particulates. For example, lead particulates in air may be determined by collecting the particulates from 0.1 to 1.0 m^3 of sample. The sample spot is then centered exactly on the surface of the ring oven and the lead dissolved by means of ammonium acetate. The lead, as the acetate, is then washed to the ring zone where it deposits as a sharply defined ring. After proper conditioning of the deposit to eliminate possible interferences, the lead is determined by reaction with dithizone and comparison of the red stain of the lead dithizonate with standard rings.

The success of the ring oven approach depends on the reliability of the final colorimetric test reactions coupled with the conditioning steps required for the elimination of interferences. Because all air samples will contain iron and lead, years of research were devoted to develop a procedure whereby these pollutants could be tolerated without danger of their invalidating the determination of other trace metals. The elegance of the ring oven approach to the study of airborne particulates therefore hinges on the simplicity of the combined sequential tape sampling of the air and the direct processing of the collected particulates. The procedure is as follows:

Sampling.

A. Collect 0.5-2.0 m^3 sample (depending on level of dust burden) on sample tape. Center spot on ring oven (use the 33 mm ring if a 1 inch dust spot was collected, or use the 22 mm for a ½ inch spot).

B. Add 30 µl NH_4Ac (15%); wash to ring zone with H_2O.

C. Add 30 µl NH_4Ac (15%) + 30 µl KCN (0.5%). Wash to ring zone with H_2O.

D. Remove the tape from ring oven and cut the ring into as many sectors as there are tests to be run.

The above procedure is deceivingly simple. However, it serves the critically important function of eliminating the interference of iron and at the same time solubilizing the metal species of interest. The iron is converted to the form of crystalline iron basic acetate and thus fixed in place. Lead and beryllium are solubilized as the pseudo salt and complex, respectively. The remaining metals are converted to soluble cyano complexes. The soluble species are all washed to the ring zone and are deposited there free of interfering iron salts. The samples are finally analyzed by appropriate test procedures as outlined below. Quantification is achieved by matching test rings (or sectors) against standard rings that have been prepared using standard solutions of the respective metals and conducting the test procedure in the same manner as that used for the sample unknown.

Beryllium (8).

1. Position the test sector on the ring oven with the sector about 1 mm from the heated surface.
2. Add 30 µl of 0.1M EDTA and wash to the ring with distilled water. Dry it.
3. Place the dried sector on the oven surface and apply the reagent crayon (morin).
4. Soak sector in a bath (1:1 concentration NH_4OH in methanol) for about 5 minutes. Dry the sector in a current of warm air. Examine under a UV lamp. Compare against standards.

Note: Beryllium reacts with morin to produce bright yellow fluorescent rings. The potential interferences are masked by the NH_4OH solution. The reagent crayon is prepared by adding 3.5 g glyceryl stearate to 1.5 g paraffin, heating over boiling water until transparent, adding 50 mg of morin (K & K laboratories) and finally pouring the melt into waxed-paper straws.

Limit of identification: 0.01 µg

Optimum range: 0.03 µg - 0.2 µg

Beryllium is seldom found in urban atmospheres in detectable amounts.

Cadmium (9).

1. The test sector is positioned on the ring oven (1 mm from the heated surface).
2. Add 30 μl of 10% Na_3PO_4 and wash it to the ring.
3. Dry the sector and expose it to NH_3 fumes.
4. Add the reagent (ferrous dipyridyl iodide*) to the ring using a capillary pipet. Wash in a jet of distilled water for 30 seconds. Dry in a current of warm air. Compare against standards.

Note: Phosphate masks the interference of Pb, and NH_3 masks the interference of Cu, Co, Ni, Zn, etc. Cadmium reacts specifically producing a bright pink stain. Preparation of ferrous dipyridyl iodide: 0.25 gms α,α'- dipyridyl and 0.146 gms $FeSO_4 \cdot H_2O$ are dissolved in 50 ml water, 10 gms KI are added, and after shaking vigorously for 30 minutes the solution is filtered. The $[Fe(\alpha,\alpha'\text{-}dip)_3]I_2$ solution contains excess iodide ions to accomplish the formation of the $[CdI_4]^{-2}$ ions essential to the cadmium test. The reagent solution is stable for 5 days.
Limit of identification: 0.05 μg
Optimum range: 0.08 μg - 2.0 μg
Cadmium in urban atmospheres is seldom encountered in detectable quantities.

Chromium (10).

1. Apply a blank protective crayon (3.5 gms glyceryl stearate + 1.5 gm paraffin wax, melted and poured into a straw) to the test sector.
2. Dip the sector into a bath of 0.2 N H_2SO_4 and then into a bath of a saturated solution of diphenylcarbazide in acetone. Dry the sector and compare against standards.

Note: Only Cr^{VI} (ie., CrO_4^{--}) reacts with the reagent under these conditions. The violet test stain is stable for only about 30 minutes, because of the acid-catalyzed oxidation of the reagent to a brown product. The protective crayon serves to stabilize the complex on the ring. Permanent standards are preferred. For determining total Cr (ie., Cr^{VI} and Cr^{III}), the sector is positioned on the ring oven, and 15 μl of a dilute solution of NaOBr are added. Wash to the ring. Dry the sector and treat with protective crayon. Dip into a bath of 1N H_2SO_4. When the liberation of Br_2 from the ring ceases, dip the sector into a saturated solution of diphenyl carbazide in acetone. Compare against standards.
Limit of identification: 0.1 μg
Optimum range: 0.6 μg - 2.0 μg
Chromium in urban atmospheres is rarely encountered in detectable quantities.

Cobalt (11).

1. Expose test sector to NH_3 fumes.
2. Spray with Na_2HPO_4 (5%).

in a current of warm air. Compare against standards.
Note: Lead reacts with dithizone (diphenylthiocarbazone) to produce a red stain. The interferences are converted to soluble complex species and, together with the excess reagent, are washed off by the above treatment. Cobalt interferes. In case Co is present, the sector is placed on the ring oven and 2% $NaNO_2$ is washed to the ring and then the sector is subjected to the same treatment as above. Cobalt is masked with NO_2^- through formation of hexanitritocobaltate (III).
Limit of identification: 0.1 μg
Optimum range: 0.4 - 5.0 μg
The method is specific, potential interferences are masked by the ammonaical cyanide solution. The usual lead concentration for urban atmosphere is 0.8 $\mu g/m^3$.

Nickel (14).

1. Expose test sector to formaldehyde fumes for 2 minutes.
2. Spray sector with dimethylglyoxime (1% in ethanol).
3. Expose to NH_3 fumes. Compare against standards.

Note: A brilliant red precipitate gives a specific measure of the presence of Ni. The HCHO reacts with the tetracyanonickelate complex (formed during the preliminary treatment of the sample spot) to form the cyanhydrin, thus demasking the nickel and per-

mitting its reaction with dimethylglyoxime.
Limit of identification: 0.08 μg
Optimum range: 0.1 - 1.0 μg
The average concentration of Ni in urban atmospheres is about 0.03 μg/m³.

Zinc (15).

1. Place the test sector on the ring oven.
2. Add 15 μl of NH_4Ac-acetic acid solution adjusted to pH 5.
3. Add 15 μl of 5% $(NH_4)_2S_2O_3$ solution and wash to the ring with NH_4Ac-acetic acid solution.
4. Spray the sector with reagent (0.05% dithizone in CCl_4). Compare against standards.

Note: Only Zn reacts with dithizone under the conditions specified, giving a reddish-violet stain.
Limit of identification: 0.05 μg
Optimum range: 0.1 - 1.5 μg
The zinc concentration for most urban atmospheres averages about 0.7 μg/m³.

ACCURACY

In general, the procedures described above, as applied directly to the sample tapes, can be considered to have about the same accuracies as emission spectroscopy. Errors should be in the range of 10-20% for most analyses performed by reasonably experienced analysts.

REFERENCES

1. Shubik, P., Clay, D.B., and Terracini, B., "The Quantification of Environmental Carcinogens", International Union Against Cancer, Tech. Rpt. Series, Vol 4., Geneva (1970)

2. Anon, "Trace Metals: Unknown, Unseen Pollution Threat", Chem. and Eng. News, July 19, 1971. See also Chem. and Eng. News July 12, 1971

3. Rosenfeld, I., and Beath, O.A., "Selenium", Academic Press New York (1964)

4. Monkman, L., Personal Communication

5. Dharmarajan, V., and West, P.W., Unpublished study

6. Sachdev, S.L., and West, P.W., Env. Science and Tech., 4, 749 (1970)

7. Weisz, H., Mikrochimica Acta, 1954, 140. See also, Weisz, H., Microanalysis by the Ring Oven Technique, 2nd ed. Pergamon Press, Oxford (1970)

8. Jungreis, E., and West, P.W., Anal. Chim. Acta, 44, 440 (1969)

9. Dharmarajan, V., and West, P.W., Anal. Chim. Acta, In Press

10. Dharmarajan, V., and West, P.W., Anal. CHim. Acta, In Press

11. Dharmarajan, V., and West, P.W., Unpublished study

12. West, P.W., and Pitombo, L.R.M., Anal. Chim. Acta, 37, 374 (1967)

13. Shendrikar, A.D., and West, P.W., Anal. Chim. Acta, In Press

14. Dharmarajan, V., and West, P.W., Unpublished study

15. Dharmarajan, V., and West, P.W., Unpublished study

THE ACTIVITIES OF THE INTERSOCIETY COMMITTEE ON MANUAL OF METHODS FOR AMBIENT AIR SAMPLING AND ANALYSIS

Arthur C. Stern

University of North Carolina

Chapel Hill, North Carolina

The Intersociety Committee on Manual of Methods for Ambient Air Sampling and Analysis is a collaborative effort of eight American professional societies; Air Pollution Control Association, American Chemical Society, American Conference of Governmental Industrial Hygienists, American Industrial Hygiene Association, American Public Health Association, American Society for Testing and Materials, American Society of Mechanical Engineers and Association of Official Analytical Chemists. Its main objective is to produce standardized methods of sampling and analysis for routine investigations of air pollution. It is financially supported by the Air Pollution Control Office (APCO) of the Environmental Protection Agency (EPA). Its administrative functions are handled by the American Public Health Association and its affairs are directed by an Executive Committee, consisting of one member from each of the eight constituent societies, which serves as a policy-making and editorial board.

ORGANIZATION

The work of the Committee is conducted by Substance Subcommittees. Nine of these subcommittees are responsible for preparing precise statements of procedures for methods in various substance categories (Table I). A tenth subcommittee is responsible for sections of the manual, such as sampling techniques and laboratory precautions, which are common to all methods. Each subcommittee has an expert representative from every one of the participating societies. Each subcommittee selects its own chairman. The Subcommittee Chairmen regularly meet with the Executive Committee to conduct the business of the Intersociety Committee. The Executive

Table I

Substance Subcommittees

Committee Number	Area of Responsibility
1	Sulfur Compounds
2	Halogens
3	Oxidants and Nitrogen Compounds
4	Carbon Compounds
5	Hydrocarbons
6	Metals I - Lead, Zinc, Cadmium, Iron
7	Metals II - Other Metals
8	Radioactive Substances
9	Laboratory Techniques and Precautions
10	Particulate Matter

Committee plus the ten subcommittees totally involve 88 unpaid volunteer persons. In addition, the Committee employs, part-time, an Editor, an Executive Secretary, and secretaries to each of them and to the Chairman of the Committee.

The headquarters of the Intersociety Committee and its Executive Secretary, Dr. George Kupchik, are located at the offices of the American Public Health Association in New York, New York. Its Editor, Dr. Morris Katz, is located at York University in Toronto, Ontario, Canada and its Chairman is located at the University of North Carolina at Chapel Hill, North Carolina. Dr. Katz is the second person who has been retained by the Committee as its Editor. The first Editor was Dr. Moyer Thomas, who retired from the position in 1968. The Committee has had two chairmen before the present incumbent; namely Dr. Leonard Greenburg (1963-66) and Dr. E.R. Hendrickson (1966-69). The term of the incumbent Chairman ends in 1972, at which time a new Chairman will be elected.

Coordination between the Intersociety Committee and the Air Pollution Control Office, EPA, is maintained by a Coordinating Committee comprising two representatives of each of the two organizations, under the chairmanship of one of the APCO representatives. The present Chairman is Dr. Aubrey P. Altshuller. The other APCO member is Thomas W. Stanley, Chairman of the Standardization Advisory Committee (SAC) of APCO. The ISC members are Dr. Allen D. Brandt and the author. The principal areas of coordination are in (a) the establishment of priorities for the need for standard methods and (b) the establisment of means for collaborative testing of "tentative" methods to establish their acceptability as "standard" methods.

ROLE OF CONSTITUENT SOCIETIES

Adherence to the Intersociety Committee does not strip from its constituents any of their prerogative. Therefore the intrasociety activities of the several societies related to methods for ambient air sampling and analysis have remained unaffected by their adherence to the Intersociety Committee. However, each of the constituent societies has formally endorsed the published Intersociety Committee methods and several of them also have arranged for publication in their own media of methods, abstracts of them or both.

The Intersociety Committee went for a long time (1963-69) without visible output and people began to despair of its ever becoming a viable organization. However it is now in a vigorous stage of production and publication of methods. After a long period of growing up, the Intersociety Committee has at last started to fulfill the hopes and aspirations of its founders to provide America with standards for measurement of air pollution.

COLLABORATIVE TESTING

"Tentative" methods will remain tentative until satisfactory completion of collaborative testing indicates that their category can be changed from "tentative" to "standard" method. Two collaborative testing programs are currently underway. One is being conducted by Southwest Research Institute, Houston, Texas under a contract from APCO. Since APCO is supporting both ISC and this collaborative testing program, it is anticipated that the methods tested will allow judgements to be made on conversion of ISC methods from "tentative" to "standard" method category.

The other is "Project Threshold" of the American Society for Testing and Materials (ASTM) - a member organization of ISC. "Project Threshold" has received sufficient financial support from several trade associations and industrial corporations to allow ASTM to contract with six research organizations; Battelle Memorial Institute, Research Triangle Institute, Midwest Research Institute, Walden Research Corp., Arthur D. Little, Inc. and George D. Clayton and Associates, Inc. to collaboratively test ASTM methods. ASTM was a pioneer in the publication of methods for ambient air sampling and analysis, having started in this work long before the formation of ISC and the adherence of ASTM to ISC. As previously noted, adherence to ISC does not strip from its constituents any of their prerogatives. It is therefore both right and proper for ASTM to collaboratively test its methods. However, some of the present ASTM methods are relatively old, whereas the ISC versions, which ASTM representatives helped develop and with which ASTM has concurred, have, in general, been improved by the years of experience with them since their adoption as ASTM standards. It is therefore hoped that before ASTM collaboratively tests any method fitting this description, the ASTM method be updated to conform to the corresponding ISC method. If this is done, the collaborative test of the ASTM method will at the same time collaboratively test the corresponding ISC method. If this is not done, collaboratively testing old ASTM methods that differ will add to rather than reduce confusion.

Because of this and because collaborative testing is a slow process, only a few Intersociety Committee "tentative" methods are likely to become "standard" methods over the next several years. Some methods will undoubtedly remain as "tentative" methods for some years to come for want of implementation of their collaborative testing. In fact, some "tentative" methods may, over the years, go through one or more revisions as "tentative" methods before receiving "standard" method status.

ROLE OF SUBSTANCE SUBCOMMITTEES

The procedure for developing an Intersociety Committee method can be described as follows:

Priorities for work on methods is assigned to Substance

Subcommittees by the Executive Committee on advice of the Intersociety Committee-Air Pollution Control Office Coordinating Committee. Such assignment is not a command to produce a publishable method, since the Substance Subcommittee can report the absence of an acceptable method and its unwillingness to offer a method for approval as a "tentative" method. Except for such assigned priorities, each Substance Subcommittee develops its own schedule of priorities and assigns work on methods in the area of responsibility to its members. In general, one Subcommittee member will be assigned the task of producing an initial write-up of a method. This write-up will be reviewed and revised by the entire Subcommittee until it is acceptable to them. About 50 methods are currently in this status in committee. Subcommittees are authorized to set up task groups and use consultants from beyond their subcommittee membership to accomplish their overall task.

At this point, it must be emphasized that neither the Intersociety Committee nor its Substance Subcommittees develop or improve methods. Research and development of methods is done by individual scientists in their own laboratories utilizing financial support from other than the Intersociety Committee. If any methodological question arises during committee deliberation, it can be resolved only by the willingness of one or more committee members to see to it that the necessary additional laboratory work is done to resolve the question. Where no one can be found willing to do this, the committee has no alternative but to table the method.

ROLE OF EDITOR

When a Substance Subcommittee is ready to recommend a method for adoption as a "tentative" method, its first step is to send it to the Editor. The Editor may, if he sees fit, try to run the method in his laboratory. If, as the result of this laboratory work or his general knowledge of the method he wishes to make changes in it, he can. If the changes he requires are substantive, he sends the method back to the Subcommittee for rework and there is repetitive exchange between the Editor and the Subcommittee until the Editor is satisfied with the substantive content of the proposed method. The Editor is empowered to make editorial changes to insure uniformity of editorial format, and improve intelligibility without referral back to the Substance Subcommittee.

ROLE OF EXECUTIVE COMMITTEE

When the Editor is satisfied with the write-up of a method and has made such editorial changes as he sees fit, he transmits it to the Executive Committee with a recommendation for adoption as an Intersociety Committee method. The Executive Committee may either accept the method, accept it with a recommendation that the Editor make additional editorial changes, or may direct the Editor to return the method to the Subcommittee for substantive rework.

When the Executive Committee adopts a method, the Editor prepares it for publication and the Executive Secretary arranges for its publication, generally as one of a group of methods published in the same volume.

To date, the Intersociety Committee has not had to develop a procedure for conversion of a "Tentative" method to a "Standard" method. However, with the start of the collaborative testing program noted above, the need for such a procedure has become imminent. Therefore a new ad-hoc committee has just been set up to establish criteria for the Substance Subcommittees to use in recommending such conversion to the Executive Committee. This ad-hoc committee will have to consider the statistics of variance of data from the laboratories involved in the collaborative testing program and recommend the degree of variance tolerable in a "Standard" method.

PUBLISHED METHODS

The Intersociety Committee publishes its methods as supplements to the quarterly journal "Health Laboratory Sciences", a publication of the American Public Health Association, To date, these supplements have been to Vol 6 No. 2 (April 1969); Vol. 7 No. 1 (January 1970); Vol. 7 No. 3 (July 1970); Vol. 7 No. 4 (October 1970); Vol. 8 No. 1 (January 1971) and Vol.8 No. 2 (April 1971). These supplements are separately available from the executive Secretary of the Intersociety Committee. It is anticipated that the present policy of publication in "Health Laboratory Science" will continue. All the methods published to date are currently in the process of being consolidated into a hardcovered Manual.

A total of 49 "Tentative" methods have been published by the Intersociety Committee (Table II) and their identifying numbers are defined in Table III. The published methods include nineteen for gases and vapors (acrolein, carbon monoxide, chlorine, formaldehyde, C_1 through C_5 hydrocarbons, hydrogen sulfide, oxidant, nitrogen oxides, mercaptan, peroxyacyl nitrate and sulfur oxides); twenty-one for particulate matter (antimony, arsenic, beryllium, chlorides, dustfall, fluorides, iron, manganese, molybdenum, nitrate, polynuclear hydrocarbons, selenium, and suspended particulate matter; nine for radioactive particles (Iodine-131, lead-210, plutonium, strontium 89 and 90, tritium and gross alpha and beta radioactivity); and one for both particulate and vapor phase phenolic compounds.

INTERNATIONAL STANDARDIZATION OF AMBIENT AIR METHODS

One aspect of method standardization that is pertinent to discuss is the degree of equipment sophistication permissible in a standard method. On one extreme, a method can be developed using

Table II

Intersociety Committee Methods

11101-01-70T -	Tentative Method of Analysis for Suspended Particulate Matter in the Atmosphere (High-Vol. Method)
11104-01-69T -	Tentative Method of Analysis for Polynuclear Aromatic Hydrocarbon Content of Atmospheric Particulate Matter
11104-02-69T -	Tentative Method of Routine Analysis for Polynuclear Aromatic Hydrocarbon Content of Atmospheric Particulate Matter
11301-01-69T -	Tentative Method of Analysis for Gross Alpha Radioactivity Content of the Atmosphere
11302-01-69T -	Tentative Method of Analysis for Gross Beta Radioactivity Content of the Atmosphere
11314-01-70T -	Tentative Method of Analysis for Tritium Content of the Atmosphere
11316-01-69T -	Tentative Method of Analysis for Iodine-131 Content of the Atmosphere (Particulate Filter-Charcoal Gamma)
11322-01-70T -	Tentative Method of Analysis for Plutonium Content of Atmospheric Particulate Matter
11327-01-68T -	Tentative Method of Analysis for Radon-222 Content of the Atmosphere
11332-01-70T -	Tentative Method of Analysis for Strontium-89 Content of Atmospheric Particulate Matter
11333-01-70T -	Tentative Method of Analysis for Strontium-90 Content of Atmospheric Particulate Matter
11342-01-68T -	Tentative Method of Analysis for Lead-210 Content of the Atmosphere
12101-01-69T -	Tentative Method of Analysis for Antimony Content of the Atmosphere
12103-01-68T -	Tentative Method of Analysis for Arsenic Content of Atmospheric Particulate Matter
12105-01-70T -	Tentative Method of Analysis for Beryllium Content of Atmospheric Particulate Matter
12126-01-70T -	Tentative Method of Analysis for Iron Content of Atmospheric Particulate Matter
12126-02-70T -	Tentative Method of Analysis for Iron Content of Atmospheric Particulate Matter (Volumetric Method)

12132-01-70T - Tentative Method of Analysis for Manganese Content of Atmospheric Particulate Matter

12134-01-70T - Tentative Method of Analysis for Molybdenum Content of Atmospheric Particulate Matter

12154-01-69T - Tentative Method of Analysis for Selenium Content of Atmospheric Particulate Matter

12202-01-68T - Tentative Method of Analysis for Flouride Content of the Atmosphere and Plant Tissues (Manual Method)

12202-02-68T - Tentative Method of Analysis for Flouride Content of the Atmosphere and Plant Tissues (Semiautomated Method)

12203-01-68T - Tentative Method of Analysis for Chloride Content of the Atmosphere (Manual Method)

12306-01-70T - Tentative Method of Analysis for Nitrate in Atmospheric Particulate Matter

17223-01-69T* - Tentative Method of Chromatographic Analysis for Benzo(a) Pyrene and Benzo (k) Flouranthene in Atmospheric Particulate Matter

17242-01-69T - Tentative Method of Microanalysis for Benzo (a) Pyrene in Airborne Particulates and Source Effluents

17242-02-69T* - Tentative Method of Chromatographic Analysis for Benzo (a) Pyrene and Benzo (k) Flouranthene in Atmospheric Particulate Matter

17242-03-69T - Tentative Method of Spectrophotometric Analysis for Benzo (a) Pyrene in Atmospheric Particulate Matter

17401-01-70T - Tentative Method of Analysis for Phenolic Compounds in the Atmosphere (4-Aminoantipyrene Method)

17502-01-70T - Tentative Method of Analysis for 7H-Benz (de) Antracen-7-One and Phenalen-1-One Content of the Atmosphere (Rapid Flourimetric Method)

21101-01-70T - Tentative Method of Analysis for Dustfall from the Atmosphere

42101-01-69T - Tentative Method for Preparation of Carbon Monoxide Standard Mixture

42101-02-69T - Tentative Method of Analysis for Carbon Monoxide Content of the Atmosphere (Manual Colorimetric Method)

* Both appear in same standard

42101-03-69T - Tentative Method of Analysis for Carbon Monoxide Content of the Atmosphere (Infrared Absorption Method)

42101-04-69T - Tentative Method of Continuous Analysis for Carbon Monoxide Content of the Atmosphere (Nondispersive Infrared Method)

42215-01-70T - Tentative Method of Analysis for Free Chlorine Content of the Atmosphere (Methyl Orange Method)

42401-01-69T - Tentative Method of Analysis for Sulfur Dioxide Content of the Atmosphere (Colorimetric)

42401-02-70T - Tentative Method of Analysis for Sulfur Dioxide Content of the Atmosphere (Manual Conductimetric Method)

42402-01-70T - Tentative Method of Analysis for Hydrogen Sulfide Content of the Atmosphere

42410-01-70T - Tentative Method of Analysis for Sulfation Rate of the Atmosphere (Lead Dioxide Cylinder Method)

42602-01-68T - Tentative Method of Analysis for Nitrogen Dioxide Content of the Atmosphere (Griess-Saltzman Reaction)

42603-01-70T - Tentative Method of Analysis for Total Nitrogen Oxides as Nitrate (Phenoldisulphonic Acid Method)

43101-01-69T - Tentative Method of Analysis for C_1 through C_5 Atmospheric Hydrocarbons

43501-01-70T - Tentative Method of Analysis of Formaldehyde Content of the Atmosphere (MBTH - Colorimetric Method - Applications to Other Aldehydes)

43502-01-69T - Tentative Method of Analysis for Formaldehyde Content of the Atmosphere (Colorimetric Method)

43505-01-70T - Tentative Method of Analysis for Acrolein Content of the Atmosphere

43901-01-70T - Tentative Method of Analysis for Mercaptan Content of the Atmosphere

44101-01-69T - Tentative Method for Continuous Monitoring of Atmospheric Oxidant with Amperometric Instruments

44101-02-70T - Tentative Method of Analysis for Oxidizing Substances in the Atmosphere

44301-01-70T - Tentative Method of Analysis for Peroxyacetyl Nitrate (Gas Chromatographic Method)

Table III

Identifying Numbers

12345	01	69	T
Pollutant Identification	Chronological Order of Adoption of Method for this Pollutant	Year of Adoption of Method	Tentative

In the SAROAD (Storage and Retrieval of Air Quality Data) system[1], in the five digit identification

1	2	3	4 5
A	B	C	D

A Denotes one of 9 major classes
eg 1 - Suspended Particulate
2 - Settled Particulate
3 - Respirable Dust
4 - Gas and Vapors
etc.

AB Denotes one of 81 subclasses
eg 42 - Gas and Vapors, Inorganic
17 - Suspended Particulate, Aromatic Compounds

ABC Denotes one of 720 families
eg 426 - Gas and Vapors, Inorganic, Nitrogen Compounds
172 - Suspended Particulate, Aromatic Compounds, Polynuclear

ABCDE Denotes one of a possible 72,171 individual pollutants
eg 42606 - Gas and Vapors, Inorganic, Nitrogen Compounds, Nitrogen Dioxide
17242 - Suspended Particulate, Aromatic Compounds, Polynuclear, Benzo (a) Pyrene

[1]Fair, Donald H.; Morgan, George B.; and Zimmer, Charles E. Storage and Retrieval of Air Quality Data (SAROAD) System and Data Coding Manual, National Air Pollution Control Administration Publication No. APTD 68-8. Office of Technical Information and Publications, NAPCA August 1968, Cincinnati, Ohio, 47 p.

only laboratory glassware, some tubing and a pump. On the other extreme, a method might call for use of a piece of analytical equipment costing $100,000. The Intersociety Committee has rejected both these extremes. We have rejected the former because we do not wish to degrade American sampling and analytical practice to the lowest common denominator of the most poorly equipped laboratory. We have rejected the latter because a method requiring extremely expensive equipment would find few users among the laboratories in America actually performing such analyses. We have attempted to steer a middle course of assuming that organizations that wish to make precise measurements must be prepared to make a substantial investment in precision equipment and that this should be the norm for American laboratories making such measurements.

We recognize that while this stance may be reasonable for the United States, where the Federal government has for the last several years contributed substantial amounts of money to improve the quality of state and local laboratories, it may not be a reasonable one for international standardization. We recognize that on the world scene there are many countries where priorities and resource allocation have resulted in laboratories being required to do air sampling and analysis without benefit of costly and difficult to maintain analytical equipment. Therefore organizations such as the World Health Organization best serve their membership by adopting in their standardization efforts an analytical philosophy more in keeping with the needs of their less affluent member nations. This should point the lesson that the sampling and analytical standard methods of one country should not be adopted willy-nilly by another country or group of countries without careful consideration of their relevance.

EMISSION SAMPLING AND ANALYSIS

Until now the Intersociety Committee has confined its activities to Ambient Air Sampling and Analysis. However, the Clean Air Amendments of 1970 place a new national emphasis on the measurement of emissions into the atmosphere. The Intersociety Committee subcommittee structure and mode of operation, perfected, one might say the hard way, over the past eight years is as applicable to emission measurement as to ambient air measurement. Its principal deficiency in adapting to the standardization of emission measurements is that its membership presently includes no industrial or trade organizations. For an emission test method standard to receive national acceptance, there would have to be participation and concurrence of the affected industrial or trade organizations. This could be accomplished by structuring Emission Substance Sub-Committees differently from the present Ambient Air Substance Sub-Committees, so that such committees include not only the eight nominees of the Societies constituting the Intersociety Committee but also nominees of the principal affected industrial and trade organizations. Neither the Intersociety Committee nor the Air Pollution Control Office of EPA have taken definitive action in this regard, but discussions between them have been initiated to assess the feasibility of extending the area of standardization of the Intersociety Committee into the area of emission measurement.

THE STATUS OF SENSORY METHODS FOR AMBIENT AIR MONITORING

E. R. Hendrickson, Ph.D., P.E.

Environmental Engineering, Inc.

Gainesville, Florida

INTRODUCTION

As measured by citizen complaints and by definition of the public, odors perceived by individuals comprise one of the foremost air pollution problem areas in the United States (1). Problems which are associated with evaluating and quantifying the presence of odorants in the air comprise one of the most difficult of scientific pursuits, yet these problems must be solved if we are to attempt to bring about control of odorous emissions. It is essential to determine the acceptable level of odorants in the atmosphere as a goal to improving air quality. The variability of odor responses in individuals, the low levels at which odorants frequently elicit responses, the conversions which may take place during the transport of odorants, and the effect of mixtures of odorants all contribute to the difficulty of providing solutions.

Odor perception by humans is highly variable, thus researchers and abatement officials have long hoped for instrumental methods of sufficient sensitivity to detect the low concentration of compounds which frequently produce citizen complaints. New concentration techniques and recently developed detectors (2, 3) bring detection of low levels of odorants in the ambient air within the range of feasibility for at least some compounds. These advances are helpful, but they do not provide a complete answer because instruments measure the concentration of odorants and humans experience odors. In the final analysis, the success or failure of odorant emission controls will be determined by the acceptability of the air quality to the public. For reasons already cited, correlations between odorant concentrations and perceived odors have not been too successful.

The human sense of smell is notoriously unreliable. The variations are due to psychological and physiological factors as well as external physical factors. Little is known about the physiological basis of odor perception in humans. We have some very detailed information about the structure of the odor perception apparatus, and some data about the transmission and interpretation of signals by the brain. Some generalized observations have been made about the relationship of certain odorants to the odor perceiving organs. We do not yet know, but we should find out, what factors are involved in individual variations of perception and what is the extent of individual variations. Investigations are not yet complete on the physicochemical nature of the stimulus. Does the human receptor identify compounds, radicals, or some other characteristic of the odorant? Is his evaluation of odor intensity strictly a function of the number of identity units present? Can the presence of non-odorous compounds affect odor perception? It has been observed that age, sex, prior conditioning, and extraneous stimuli all influence our perception of odors. It has been concluded that, to a considerable extent, the response of humans to odors is a learned reaction (4, 5).

SENSORY METHODS OF ANALYSIS

Despite the dearth of knowledge about the basis of odor perception in humans, and the uncertainties of sensory methodology, a number of sensory methods are being applied to describe odors qualitatively and quantitatively. A number of references appear in the literature on the use of the human senses to measure the intensity and quality of odorants and odorant mixtures (6). Relatively few of these use the same techniques. Part of the lack of uniformity in techniques, and lack of consistency in results, is due to the state of development of the art. Application of these psychophysical methods performs a number of functions. Included among other functions are evaluation of odor source strength, determination of the acceptability of an odor, determination of the threshold of detectability, evaluation of the effectiveness of odor control procedures, and determination of the frequency of odor experience. Sensory methods which are in use may evaluate one or more of the various dimensions of an odor experience.

Lindvall (7) has indicated that a given odor sensation may generally be described in terms of four dimensions. These are the detectability, the supraliminal intensity, the quality, and the hedonic tone or acceptability. Detectability refers to the pervasiveness of an odor and represents the lowest energy level capable of exciting the receptor. Intensity is a measure of the level of concentration above some threshold. The term quality refers to those properties which distinguish one odor from another, regardless of the other dimensions. Hedonic tone is a measure of the acceptability of an odor ranging from pleasurable to unpleasant. Although acceptability is believed to be the most important dimension of an odor, most of the sensory methods were originally used for determining intensity or detectability.

In June 1970, a week-long symposium was held in Stockholm for the purpose of evaluating methods of odor detection (6). Conclusions of the conference revealed that human sensory analysis can be applied to ambient air sampling in two main ways. The first method is described as prospective and is based on sequential observations of odor in a given area for a fixed period of time. This method uses a panel of either trained or untrained observers. The second method is described as a retrospective procedure which involves an after-the-fact opinion survey of the resident population regarding odor frequency and duration. This paper is concerned only with the former type of method. No concensus could be reached at the symposium by which it would be possible to recommend a single sensory method of evaluating odors. It should be recognized, however, that most of the methods which are presently in use contain certain common elements. Included in these elements are selection of observers, presentation of the stimulus, elimination of extraneous stimuli, and evaluation of one or more dimensions of the stimulus by the observer.

Selection of Observers

It was previously indicated that substantial variations are observed in the dimensions of perceived odors from one person to another, and in a single individual from time to time. Age, sex, prior experience, learned responses, and extraneous stimuli all influence an individual's perception of odor. Thus, selection and training of subjects who are to be used as judges is of great importance. Guidelines for selection and possible training of judges will depend upon the objective of the investigation. If the purpose of the study is to evaluate the hedonic tone or the frequency of an odor experience in the environment, the panelists should be demographically representative of the entire population. If the purpose of the study is to evaluate the intensity of exposure or determine thresholds, the panelists may be selected to represent certain perceptual variables. Studies of odor quality may require selection of observers who can reproduce their perceptual judgment within a narrow range. It may or may not be desirable to provide extensive training for a panel. The desirability of training is dependent upon the objectives of a given investigation. Generally, in the type of studies under discussion, orientation of the subjects regarding the nature of the study and familiarity with test procedures helps to yield more consistent and significant information. Preliminary runs are frequently desirable.

Presentation of the Stimulus and Response

An important consideration in any sensory method is the manner in which the odor stimulus is presented to the judges. A number of devices, frequently referred to as olfactometers, have been developed and used over the years. An objective in selection of an olfactometer is to present the stimulus in as natural a manner as possible, eliminating such extraneous elements as other odors, variations in temperature and humidity, or unintentional dilution of the odor. This may be accomplished by selecting a chamber large enough to accommodate the test subjects who are completely surrounded by the odorant gas. Other workers have used an odor hood which covers only the nose, face, or head. Still another method is to introduce the odorants directly into one or both nostrils. See page 31 of reference 6 for further references. Each method has its own advantages and disadvantages. Regardless of the physical arrangement for presentation of the stimulus, it is usually desirable to dilute the odorant air with deodorized air, nitrogen, or other mixtures. Thus a mechanism is necessary to provide for static or dynamic dilution (7, 8).

The presentation of the stimulus to the subjects requires careful consideration of the purpose for which the study is being made. The Stockholm symposium previously mentioned identified three methods for presenting the stimulus when thresholds were to be determined. The first was the method of limits in which various dilutions are presented in ascending or descending series starting at different levels each time. The subject is required to report whether or not the stimulus can be detected. The threshold value for each stimulus then is defined as the mid-point between the limits of detection and no detection in the series. A second approach is described as the method of constant stimulus, where the odorant gas is presented in a random selection of intensities. At each intensity, the subjects are required to indicate only if they can detect the stimulus. The third method is based on the theory of signal detectability. The theory assumes that responses to the same repeated stimulus are normally distributed. "Noise" is assumed to be constantly present as an integral part of the stimulus. The response may be due to the presence of noise alone, or of the stimulus under consideration plus noise. In this method, a stimulus is presented in random order alternating with noise. By this means, the relationship may be obtained between correct and incorrect positive answers. In any of these methods of presentation, the response may be forced choice with as many as four alternatives, semi-forced choice with two levels of presentation and three choices, or two response categories with answers such as yes or no.

Where dimensions of the odor other than detectability are to be considered, a different approach may be used. The presentation here usually involves comparison with standards of either intensity, or quality and acceptability. The subjects may be presented with one or more stimulus levels and a standard, and asked to make a comparison. For analysis of odor quality and acceptability, he may be presented with the stimulus and one or more standards and asked to make the response that the stimulus "smells like ...". A unique response method employs a suitably drawn "cartoon type" of scale which depicts various levels of like and dislike (9).

Elimination of Extraneous Stimuli

Mention has already been made of the potential effect of extraneous stimuli on odor perception studies. It is important to eliminate as many of these as possible. Thus, application of sensory methods in the field usually requires the use of an elaborate mobile laboratory in which the temperature and humidity can be controlled, outside noises excluded, the subjects made comfortable, and the stimulus presented in the most natural manner.

CONCLUSIONS

The conclusions of the Stockholm conference were that it was premature to attempt to standardize on any single sensory method. The desirability of such standardization was recognized, however. A new seminar on odors is scheduled for Boston in late April, 1971. At this time, perhaps, at least tentative standardized procedures can be agreed upon. It is important to do this as a base for further refinement, and should be an international development. In the meantime, the guidelines for the selection of sensory methods of analysis which were developed at Stockholm should be helpful in making certain decisions:

"1) No finding is any better than the pains taken by the investigator in conducting his study, especially as to the selection of a panel method, care of the stimulus material, and instruction of the individual panelists.

2) For any study the environment itself is important. For some studies (detection) a relatively odorless area is essential prior to testing.

3) Information which might bias the panelists response must be avoided. This includes talking or other signals, unwanted identification of specimen, differences in appearance, noise and the like.

4) If the panel is to speak for the greater community every care must be taken to assure that the panel embodies a representative sampling of the key demographic characteristics of the larger population.

5) If an odorant or odorants are identified and isolated for study as typical of the unwanted source of an odor, be certain that this odorant is clearly identified from physiochemical measurement standpoint as representative of those in the atmosphere.

6) Always be certain that the panelist is clear on his task by giving explicit precise complete instructions.

7) Discourage all contamination of any measurement by panelists by preventing them from smoking, eating or drinking just before or during tests and also prohibit the use of perfumed cosmetics on his person.

8) Old people are more prone to have suffered loss in olfactory sensitivity, which should be considered in the studies where accurate sensing must be accomplished."

REFERENCES

1. _____, National Survey of the Odor Problem, Phase 1 of a Study of the Social and Economic Impact of Odors, NAPCA Contract 22-69-50, Copley International Corporation, La Jolla, California, January 1970.

2. Hanna, G.F., Odor Measurement Tools Available to the Engineer, Paper presented at the 67th Annual Meeting of the American Institute of Chemical Engineers, Atlanta, Georgia, February 15-18, 1970.

3. Dravnieks, A., Personal communication, 1971.

4. Moncrief, R.W., Odor Preferences, John Wiley, New York, 1966.

5. Engen, T. and Katz, H.I., Odor Preference and Response Bias in Young Children, Unpublished manuscript, Brown University, 1968.

6. _____, Methods for Measuring and Evaluating Odorous Air Pollutants at the Source and in the Ambient Air, Report of an International Symposium, Karolinska Institute, Stockholm, 1970.

7. Lindvall, T., "On Sensory Evaluation of Odorous Air Pollutant Intensities", Nordisk Hygienisk Tidskrift, Supplement 2, Stockholm, 1970.

8. Committee D-22, "Standard Method for Measurement of Odor in Atmospheres," (D1391-57), ASTM Standards, Part 23, November 1970.

9. Springer, K.J. and Hare, C.T., "A Field Survey to Determine Public Opinion of Diesel Engine Exhaust Odor," Report No. AR-718, Southwest Research Institute, San Antonio, 1970.

INTERFACING OF SENSORY AND ANALYTICAL MEASUREMENTS

Andrew Dravnieks

Odor Science Center, IIT Research Institute

Chicago, Illinois 60616

INTRODUCTION

The objective of odorous air pollution control is to reduce or eliminate the complaints by the affected population.

In measuring the air pollution caused by odorants, a distinction therefore must be made between the analytical description of the odorous stimulus and the sensations produced in humans by this stimulus. The analytical description reveals the odor-relevant species and their concentrations in the air sample. The sensory (psychophysical) response depends on the type of species and their concentration in a complex, sometimes only yet vaguely known way.

Nevertheless, interpretations of the analytical data in sensory terms are becoming increasingly necessary. The following article outlines the state-of-art in estimating the sensory effects of analytically measurable changes in odorant concentrations.

ODOR SENSATION

Odor sensation is produced by substances acting on the olfactory and trigeminal chemosensing systems of living beings. In humans, the olfactory sensors are located in a certain patch of each of the nasal passages; the trigeminal sensors -- bare nerve endings of the trigeminal nerve system -- are dispersed in the nose, mouth, and area close to the eye.

It is generally accepted that odorants can stimulate either of these chemosensing systems, but that the olfactory system is more articulate and in general more sensitive to more substances, while a small number of odorants significantly or predominantly stimulate the trigeminal system. It is commonly believed that the irritation caused by some chemicals in nose, throat, or eyes is an essentially trigeminal effect, and that the concentrations needed to cause irritation are usually considerably higher than those which cause the olfactory sensation. However, some substances may not exhibit much "odor" but can cause irritation leading to coughing, eye watering, and other objectionable reactions. A typical example is the reaction produced by exposure to sulfur dioxide.

ODOR-RELEVANT AIR ANALYSIS

There is little evidence that non-volatile matter in aerosol form, at concentrations encountered in air pollution, can cause the sensation of odor. Consequently, the odor-relevant analysis deals with vapors.

Only a few inorganic substances are sufficiently volatile and only a few of these exhibit odors. The principal inorganic odorants are ozone, hydrogen sulfide, ammonia, and halogens; methods for measurements of their concentration in air exist.

Most odorants are organic compounds, for which air analysis is usually based on gas chromatographic procedures. Hydrogen flame ionization detector responds to practically all organic compounds essentially in proportion to their mass. Flame-photometric sulfur detector is used for compounds containing sulfur, especially the particularly odorous low-molecular thiols and sulfides. Some detectors exist or are under development for other specific species.

Several factors must be considered in analyzing air for odorants. Most organic substances with reasonable, even if low volatility do exhibit odor when smelled in concentrated form. However, many become non-detectable by human nose after even a moderate dilution with air. Therefore, it is not useful to automatically consider all organic substances that are analytically detected in air as the odor-relevant pollutants simply because they would produce odor when smelled in pure state.

On the other hand, since many substances do produce odor at concentrations as low as one part per billion, the analysis must be sufficiently sensitive to reflect the presence and measure the concentration of all those organic substances that might be odor-relevant to at least 0.1 ppb air. In addition, the procedures must provide some discrimination as to which species are odor relevant. At present, the method which seems to satisfy these requirements best is a combination of the analytical and sensory techniques.

When air is analyzed with sufficient sensitivity, it is usual to detect 50 to 100 organic components, and a gas chromatographic analysis is a necessity to separate all possibly odor-relevant components. The occurrence of odorous compounds becomes rarer beyond the C_{18} alkane elution position in Carbowax 20M column (Kovats Index >1800); an informal poll of perfumers indicated that beyond C_{22} position in this column odorants are not observed. Thus, an analysis in similar column with programmed temperature rise which encompasses all compounds up to the C_{20} or C_{22} alkane retention positions reasonably assures the analytical detection of all possibly odor-relevant components, although other columns may be needed to produce better peaks for some specific groups of compounds.

The sensitivity requirement determines the sample size and involves matching the sensory and the analytical sensitivity levels with the routine version of hydrogen flame ionization detector. In a combination with a typical electrometer and recorder, an organic component present in the sample in an amount of $2 \cdot 10^{-9}$g typically produces a peak with an area of the order of 1 cm^2 at high sensitivity setting. Sensory detectability levels for some odorants in air can be as low as 0.2×10^{-9}g/liter.

Thus, to bridge the sensory and the analytical detectability, it is necessary to analyze the organic fraction contained in several liters, typically 10 l of air, in one single analysis. Such volume of ambient air at 25°C and 50 percent relative humidity contains 120 mg of water. If water were co-condensed with the organic compounds, only a small aliquot could be analyzed in a single analysis, and the instrumental methods have been developed to collect organic species from ten-liter air samples without large accumulation of water (1,2,3).

Two situations occur in analysis of air for odorous pollutants. In some cases, unfortunately rare, the identity of the odorants is known, and the analysis amounts to a sufficiently sensitive measurement of their concentrations by methods free of interferences. For instance, the determination of methylsulfide using a gas chromatographic process and a flame photometric sulfur detector (even if the response of this detector is non-linear) is preferable to the linearly-responsing flame ionization detector, which detects all organic compounds and may produce a peak in methylsulfide position because of response to some sulfur-free organic compound.

In most cases, the exact nature of the pollutants responsible for the odor is, as yet, unknown. In the gas chromatographic analysis of such samples, using the flame ionization detector only, a feeling of futility is inevitable when inspecting the gas chromatogram which registered the presence of very many species. Here, a sensory assay of the effluent from the gas chromatographic partition column is extremely helpful (3,4). This effluent is split in two portions; one flows to the detector, another to a sniffing port where the chemist observes the odors of the components and annotates the gas chromatogram in terms of the odor notes.

If an organic fraction collected from 10 liters of air is analyzed, the gas chromatographically separated components emerge from the sniffing port at concentrations up to two orders of magnitude higher than that initially in the air. The components carrying characteristic odor notes can be located in the gas chromatographic retention scale. Those -- usually many -- which do not exhibit odors can be tentatively ignored since apparently they have been present in air much below detectability level. Different odorous air samples are evaluated by comparing the areas of the respective characteristic odor peaks, even before the chemical identity of the components has been established. Results of such GC-sensory assay are most informative, especially where several possible odor sources may contribute to the pollution odor.

Packed or supported-coated open tubular gas chromatographic partition columns must be used since a substantial sample must be analyzed; capillary columns produce better separation, but can only accept a much smaller sample size. Separation can be enhanced by two-column chromatography discussed elsewhere (3).

ANALYTICAL DATA IN SENSORY CONTEXT

When quantitative differences are observed in the analytical composition of two samples of odorous air or exhaust, their traditional chemical interpretation is simple. For instance, sample I may contain 20 percent more of methylsulfide than sample II, and if the analytical accuracy is of the order of ± 5 percent, one might conclude that sample I is more odorous. However, for the human nose, an increase in the odorant concentration by 20 percent results in a barely noticeable odor intensity increase. These and other characteristics of human response are discussed in more detail in the next sections, and must be taken into consideration in estimating what degree of change in the sensory response can be expected from observed change in the analytical composition.

DIMENSIONS OF ODOR

The sensory (psychophysical) dimensions of the odor sensation are its intensity, detectability, acceptability, and quality (5).

The sensation can be weak, moderate or strong, or it can be characterized by reference to some other scale, including odor intensity reference kits (6,7). Detectability is the term for characterizing those concentration levels at which the odor intensity becomes low and the odorant's presence is difficult to detect.

Acceptability, in air pollution situations, is not synonymous with desirability, as with aromas, e.g., in toiletry marketing research. In air pollution, the term acceptability usually has only a negative direction -- annoyance: those odors which produce less annoyance reaction are "more acceptable;" a fragrance desirable in a perfume may result in an annoyance reaction when habitually emitted from a manufacturing plant.

Quality refers to the character of odors in terms of similarity to various known odors; of course, the acceptability is one of the aspects of quality. The initial judgement in the comparison of two odors usually deals with their relative pleasantness or unpleasantness; decision on differences or similarities in the odor character of, for example, two similarly unpleasant odors requires additional judgement.

Judgement Criteria. Engineers and chemists are accustomed to dealing with exact sciences where properties of odorants, such as density, boiling point, and other characteristics, can be measured in exact terms and are called "constants." They expect to measure "odor dimensions" just as exactly. If only, for instance, one could assemble "enough" people with "competent reproducible" judgements and could accumulate "enough" data, one should be able to "pin down" the odor threshold value to three significant digits.

An engineer is dissatisfied because odor threshold values are available for only a limited number of compounds, and that various authors report values sometimes differing by a factor of ten or more. It is somewhat easier to accept that the annoyance reactions of people may differ.

The recent advances in psychophysics, including those based on introduction of the concepts of signal detection and decision theories, lead to a more sophisticated attitude toward the "irreproducibility" of sensory data. Thus, the notion of a threshold as an exact characteristic is eroding. In essence, in each judgement, the subject weighs the possibility that only spurious sensations ("noise") are present vs. the possibility that an odor or a "signal" is present together with the "noise." Noise here does not mean acoustic noise, but all kinds of influences, outside and inside the subject, that may mistakenly lead him to conclude that an odor is present.

Even if only a simple "yes" or "no" response is asked, the subject really weighs various possibilities: for example,

Very sure there is an odor.

Somewhat sure there is an odor.

Not sure about either the presence or absence of an odor.

Somewhat sure there is no odor.

Very sure there is no odor.

The subject then applies a criterion for setting the boundary between his "yes" and "no." The criterion is a decision parameter. It is a property of the subject, and it depends on the consequences of his judgement. Economic, social, and prestige factors are involved. The possible judgements are classified as follows:

Odor present and reported ("hit").

Odor present but not reported ("miss").

Odor absent but reported ("false alarm").

Odor absent and reported to be absent ("correct judgement").

Only two of the above possibilities can vary independently. It is common to use the ratio of probabilities of (hit):(false alarm) to characterize the response criteria. The occurrence of hits - correct detections -- can be increased only at the expense of tolerating increased frequencies of false alarms. The criterion may vary; if there is a reward for "hits" but no penalty for "false alarms" -- one obviously may be tempted to always respond, "yes, odor present." However, for example, if the subject thinks he will not qualify for rewards if too many false alarms accumulate, his criterion will shift toward a judiciously exercised frequency of false alarms and perhaps misses as well. In general, the response of a subject in terms of the (hit):(false alarm) probability shifts with criteria and can be described by so-called receiver operating curves, rather than by some well-defined "threshold" points. Similar concepts apply to the intensity, odor similarity, and acceptability judgements. The important point is that criteria-free judgements are simply impossible, and applying response criteria to odor sensations is an integral part of the sensory measurements.

Panels. In the light of these concepts, a selected, well-trained panel is simply a sensory instrument which is able to maintain a constant criterion, but its judgements are poor indicators of the response to be expected from a broader section of population where diverse and changing criteria are factors.

Such panels are most useful and needed in discrimination tests to establish to what degree and in which sensory direction a sample differs from a target sample (8). However, it has been equally accepted that such panels do not predict market response, and that the acceptability is best measurable by using many untrained individuals, who indicate the response range much more adequately. In studying odorous air pollution, the possibility of strong disagreement between highly trained subjects and the population would result from the outlined criteria variables, even if the sensory organs of all people had exactly the same sensitivity and selectivity. The complicating factor is that population, exposed to certain pollution odors, automatically becomes "trained" by learning to recognize the familiar odor.

The essence of the above discussion is that disagreements in the sensory judgements are natural and should be accepted as an integral part of the sensory data. The "average value" of, for example, detectability must be considered jointly with the distribution of detectability among different people with different physiological sensitivities and different judgement criteria.

Trying to define odor properties beyond a certain range is futile. Carefully controlled experiments with selected, highly trained subjects simply eliminate some variability from the responses. Such experiments result in knowing more and more precisely the controlled responses which, however, simultaneously become less and less meaningful in obtaining a perspective on the expected response of the population.

With this understanding, current knowledge on the functions relating the sensory and analytical data can be discussed.

SENSORY RESPONSES TO ANALYTICAL VARIABLES

Odor Intensity. This factor changes with the odorant concentration in reasonable agreement with an expression (9):

$$I = kC^n$$

where I is the perceived odor intensity; C, the odorant concentration in air; k is a proportionality factor; and n is a fractional number, which varies with odorant and may be in the range between 0.2 and 0.8. The value of n for an odorant or a mixture can be established by presenting various concentrations of the odorant and asking the subjects to give relative numerical magnitude estimates; for example, the subject is told to consider the odor intensity of a certain odorant concentration equal to 10, and to give proportional numbers to the odor intensities of other concentrations -- lower than 10 if the odor is weaker, and larger than 10 if the odor is stronger. Care must be taken to minimize olfactory fatigue (adaptation) effects. The above equation results in a straight line plot of log I vs. log C. Figure 1 shows plots obtained by such procedure.

The exponent for n-butanol has been established by us and elsewhere (10). A value of n = 0.63 tentatively appears to be a satisfactory compromise. Thus, an n-butanol reference scale (7), consisting of 8 binary dilution steps can be used and other odorants at several

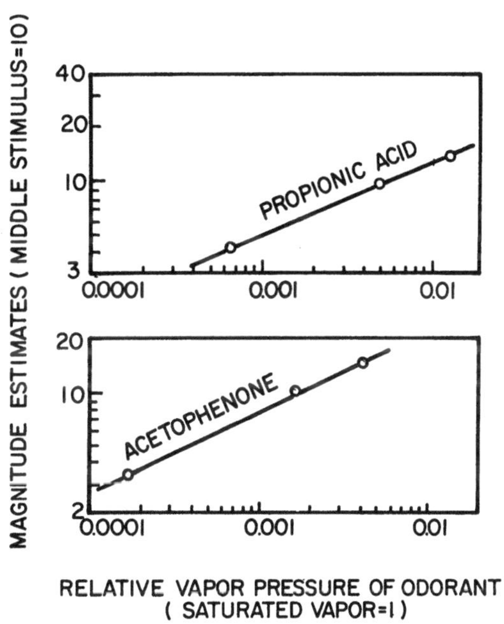

Figure 1
TWO EXAMPLES OF NUMERICAL MAGNITUDE ESTIMATES OF RELATIVE ODOR INTENSITIES

The plots are not related to each other: propionic acid and acetophenone at intensity 10 here do not smell equally intense. Exponents n: propionic acid, 0.42; acetophenone, 0.45

concentrations matched against this scale, disregarding the differences in odor quality. With a minimum interference by the adaptation effects, log I vs. log C plots for other odorants and odorous effluents have been repeatedly obtained in our work using this scale.

Such intensity rating procedure differs from the category assignments, where odors are rated in, for example, 5 categories, such as 1 = barely perceptible, 2 = faint, 3 = easily noticeable, 4 = strong, 5 = very strong (11). These numbers are category designations and are not proportional to the odor intensities: for instance, an odor of category 4 in actuality seems 3 to 4 times stronger than that of category 3. Of course, if the odor intensity ratios for a specific category scale have been worked out, one can relate the category-rated odor intensities to the values of plots of Figure 1. Figure 2 gives odor intensity plots for several odorants vs. n-butanol scale, annotated by numbers proportional to the respective odor intensities, with the intensity 10 assigned to an "average" (our work) odor intensity. Katz-Talbert categories, (11) are also indicated.

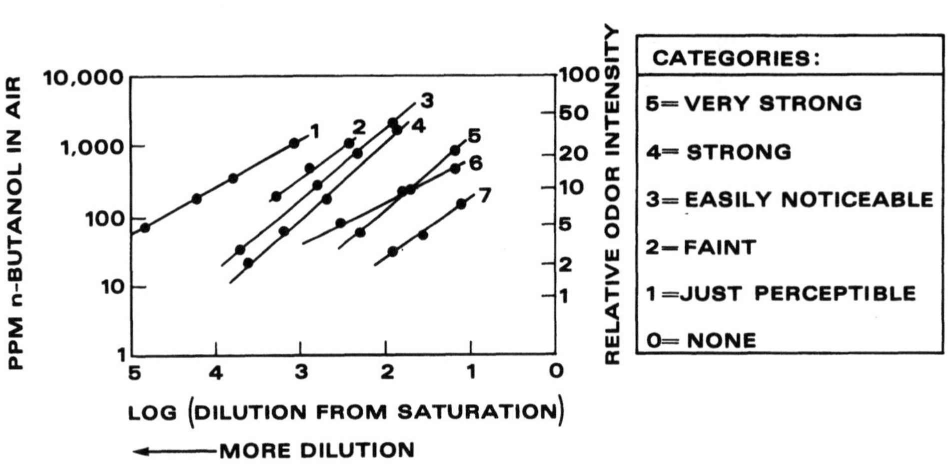

Figure 2
ODOR INTENSITY VS. ODORANT CONCENTRATION
Left Scale: ppm n-butanol exhibiting respective odor intensities
Right Scle: author's relative odor intensity scale, and Katz-Talbert category scale
1, Thiophene; 2, pyridine; 3, n-hexanal; 4, allylalcohol; 5, benzene; 6, propionic acid; 7, isopropanol

The consequence of these correlations is that a major decrease -- by a factor of 4 to 8 -- in the odorant concentration by an emission control is needed to decrease the "felt" odor intensity by a factor of 2. Such decrease will be barely noticeable unless the two odor stimuli are directly compared. Thus, an analytically most-obvious decrease in the concentration of the odorant by a factor of 8, for an odorant with $n = 0.33$, may not impress subjects as a significant odor reduction.

If two odorant concentrations can be smelled side by side, it becomes difficult to establish which is stronger if the concentrations differ by less than 20 or 30 percent (differential threshold). Accuracy of the analytical methods usually is considerably better.

Detectability. When, as a result of a reduction of the odorant concentration by some effluent treatment, the odor intensity becomes low, a threshold range is reached where the detection of the odor becomes difficult. As discussed in the section dealing with odor dimensions, a "magic" value of "threshold" cannot really be established. Even the pre-signal detection-theory version of the threshold definition -- that concentration below which the odor is not detected in 50 percent of trials -- usually is not suitable in odorous pollution studies. Individuals have many thousands of trials per day, and a detection of an odor in a few of these is sufficient to complain about odor; neither a statistical nor an engineering argument will help.

The real question remains: Is concentration of the odorous effluent low enough to eliminate the complaints or reduce them to a tolerable level? The exact answer is difficult, but plots of detectability vs. concentration are useful. Such plots must be generated by using subjects who are diverse in sensitivity and criteria; if they are "selected" at all, they must represent the distribution of a larger segment of population. Figure 3 is an example drawn on a log probability scale. Extrapolation to a few percent ordinate gives an estimate of levels at which only a small fraction of individuals may still detect an odor. These low-detectability levels, rather than the 50 percent response point, are more realistic criteria for the control of odors in the ambient air.

Figure 3 was obtained in a sensory laboratory with a panel composed of individuals with low, medium and high

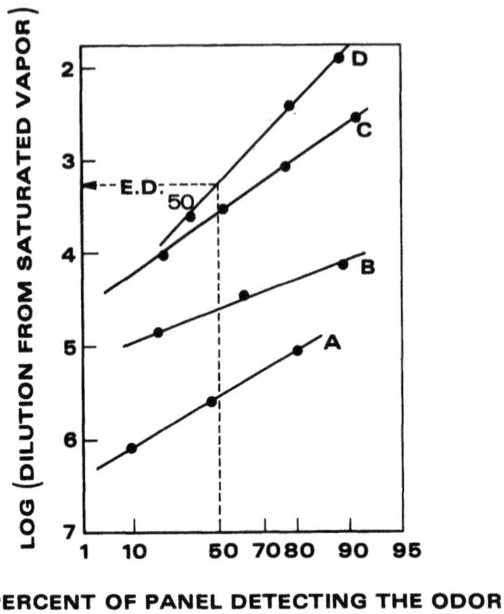

Figure 3
DETECTABILITY PLOTS
A, Butylacetate; B, 2-Me-2-Propanol;
C, Toluene; D, Isopropanol

sensitivity, selected from a larger group of more than 60.

An alternate method is to use the residents of the community affected by the odorous pollution, under the normal community conditions, as a response group, ("panel at large"), and conduct ambient air analyses at different levels of complaints (to characterize the odor stimuli analytically). Such procedures were used in a village next to a resin reactor. Only a limited number of inhabitants were particularly sensitive or lived in the particularly affected locations. Air samplings were conducted at locations and times of complaints as well as at nearby locations where the same people detected no odor. It was possible to establish that the complaints, within the demography of the village, ceased when the effluent in the ambient air became diluted by a factor of 10^6.

To summarize, curves indicating the change of detectability with the odorant concentration are useful tools for engineering calculations of odorous air pollution control. Even if only approximately obtained, they are functionally more useful for estimating the response than the classically construed odor threshold value.

Because of the differences in the slopes of odor intensity vs. concentration curves, the measurement of the odorous effluent intensity in odor threshold units also has a limited utility, although it has been a time-honored custom. Threshold multiples as estimates of the as-is odor intensities are certainly invalid, and they beg the question: "which threshold?" Unless the dilution grossly exceeds the factor obtained by osmoscopes or other devices anchored to a "normal" or "skilled" subject and employing basically inadequate statistics, the odor control by effluent reduction or dilution will fall far short of the level where complaints actually can be expected to cease.

The contemporary daily press is full of anecdotes about areas where odor presumably "should not exist" but complaints continue. The analytical specifications for reaching specific no-complaint levels are better established by departing from the conventional threshold concept. In turn, the analytical data on some actual air sample should permit at least crude estimates of the expected complaint levels.

Acceptability (Quality). These two odor dimensions are closely related to each other and to the nature of odorants and their concentrations in air.

One must distinguish between the acceptability (pleasantness-unpleasantness) in absence of specific context when chemicals are smelled in a sensory laboratory, and the acceptability within a definite context or expectation. In laboratory, there is usually little disagreement about some odorants recognized to be generally unpleasant at higher concentrations; more disagreement exists on what is pleasant odor, and considerable diversity of opinion occurs about those odors which fall into categories of somewhat pleasant, neutral, or somewhat unpleasant.

Figure 4 gives a hedonic rating by the same panel of three odorants at a series of concentrations. Intensity has a secondary effect. Propionic acid was pleasant to some at low concentrations, unpleasant at higher; phenylethylether with an aroma described as rose-like, was "unpleasantly sweet" to some subjects.

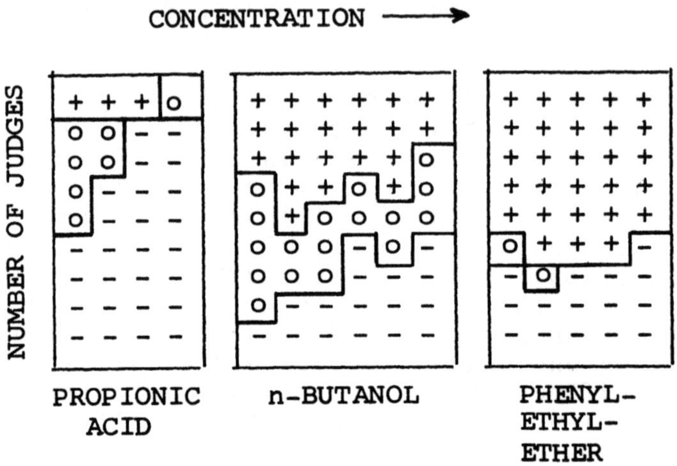

Figure 4
HEDONIC CONCENTRATION PROFILES
Each column represents the odorant concentration 3 x higher (for n-butanol, 2 x higher) than the adjoining column on its left.
(+) = pleasant; (-) = unpleasant; (o) = neither

Environmental Perplexities. In environmental contexts, expectations greatly influence judgements. Odors are frequently assigned to sources on the basis of various preconceived notions. Here, the analytical data on the various effluents in the particular community are extremely helpful in unravelling the situation, especially if supplemented with the detectability estimates of the various odorants.

If the analytical data indicate high-detectability levels of source A, for example, a garbage dump, and low-detectability levels of source B, for example, a non-smoking stack, one must expect that if some subjects attribute the odor to the visible source B, it is unlikely that the complaints will be fully eliminated by controlling the effluent A. Elimination of the visible source B can occasionally reduce the complaints about odor which actually continues to originate from A. The judgement criteria depend not only on the analytical data in air, but also on the whole set of contextual expectations. Even if the effluent reduction from source A may not completely eliminate complaint as long as the visible stack B exists, the analytical data will at least show that such complaints are in the "false alarm" category and a public education program will be indicated.

CONCLUSIONS

Analysis obtained with sufficient sensitivity and sufficient resolution into chemical species provides "chemical photographs" of odor stimuli.

The human response to the changes in the analytical composition depends on the context of the stimuli, the physiological sensitivities of the subject, and the judgement criteria. These criteria are influenced by social, economic, and prestige considerations of the consequences of specific judgements. The variability of human response is an integral part of sensory judgements and requires use of broadly-selected panels in contrast to selected trained panels showing agreements in response.

Some sensory/analytical relations, for example, change of odor intensity with concentration of odorant, are sufficiently well known, or can be obtained for various effluents, to grossly estimate the response expected from certain analytical changes. In other response effects, annoyance reaction, for instance, the relevant factors, such as the contextual influences, are recognized, but their effect cannot yet be estimated.

REFERENCES

1. Dravnieks, A. and Whitfield, J., Gas chromatographic study of air quality in schools, ASHRAE Transactions 1971, in press.

2. Dravnicks, A., Krotoszynski, B.K., Whitfield, J., and O'Donnell, A., High-speed collection of organic vapors from the atmosphere, submitted to Envir. Sci. & Technology.

3. Dravnieks, A. and O'Donnell, A., Principles and some techniques of headspace analysis, J. Agr. Food Sci. 1971, in press.

4. Guadagni, D.G., Requirements for coordination of instrumental and sensory techniques, in ASTM STP 440 Philadelphia, 1968.

5. Methods for Measuring and Evaluating Odorous Air Pollutants at the Source and in Ambient Air, 3rd Karolinska Inst. Symp. Environm. Health, Stockholm, 1970.

6. Turk, H., Diesel exhaust odor evaluation, U.S. Dept. Health, Educ. & Welf., Env. Health Series, Air Poll. Publ. 999-AP-32 (1967).

7. Perez, J.M. and Landen, E.W., SAE Preprint 680421 (1968).

8. Manual on Sensory Testing Methods, ASTM STP 434, ASTM, Philadelphia (1968).

9. Stevens, S.S., in Sensory Communications, Rosenblith, ed., p. 1, MIT Press, Cambridge, Mass (1961).

10. Cain, W.S., Perception & Psychophysics 6, 349 (1969).

11. Katz, S.H. and Talbert, E.J., U.S. Dept. Commerce, Bureau of Mines Techn. Paper 480 (1930).

STUDIES OF SULFUR COMPOUNDS ADSORBED ON SMOKE PARTICLES AND OTHER SOLIDS BY PHOTOELECTRON SPECTROSCOPY*

L. D. Hulett, T. A. Carlson, B. R. Fish; J. L. Durham

Oak Ridge National Laboratory

National Air Pollution Control Agency

ABSTRACT

Photoelectron spectroscopy as a means of air pollution monitoring and as a tool for smoke pollution research has been studied. Field samples of fly ash from a power plant and smoke particles from a home furnace have been analyzed. It is shown that the oxidation states of sulfur in compounds adsorbed on these solids can be determined. The use of photoelectron spectroscopy for studying the effects of combustion variables and downwind conditions on the oxidation state of sulfur on fly ash and smoke particles is discussed. The composition of the solid on which SO_2 is adsorbed was found to be an important variable in the rate at which oxidation to sulfate occurs.

INTRODUCTION

Air pollution monitoring and research usually involve the study of gases and particulate matters as separate components. It is important, however, to study also the interaction between the two phases. It is known that aggressive gaseous compounds are adsorbed on the solid surfaces, and in this form they may be more dangerous than in the gas phase. For example, small particles penetrate the pulmonary regions more than reactive gases. Their adsorbed chemicals are kept with them instead of being expelled, as are gases. For monitoring procedures to be more complete, the capability of in situ analysis

*Research sponsored by the U. S. Atomic Energy Commission under contract with Union Carbide Corporation.

of the surfaces of solid particles is needed. The design of effective abatement processes for gaseous pollutants such as SO_2 also requires a knowledge of gas-solid interactions. There are several abatement schemes under study which involve adsorption of SO_2 on solids. Their efficiencies depend on such factors as the oxidation state of the sulfur after it is adsorbed.

Photoelectron spectroscopy, sometimes called ESCA (Electron Spectroscopy for Chemical Analysis[1]), is a useful means of studying solid surfaces. Quantities of adsorbed material as small as a monolayer can be detected, and oxidation states of elements can be determined. Measurements are made in situ on the solid surface, and they are relatively quick compared to other methods that might be used. This paper is a report of the use of this new technique in studying solid gas reactions important to pollution research.

PHOTOELECTRON SPECTROSCOPY

Instrumentation and Principle of the Method

Figure 1 is a schematic diagram of the spectrometer used in this work. Principal parts are: Specimen (A), x-ray source (B), slits (C), resolution baffels (D), analyzer volume (E), exit slit (F), and detector (G). The specimens were mounted on small plates and placed in close proximity to the x-ray source and entrance slit. As the specimen was irradiated, electrons were ejected, some of them passing through the slit. A specially shaped magnetic field, generally perpendicular to the plane of the analyzer volume, focused the electrons on the detector. The strength of the magnetic field required to focus the electrons was a measure of their kinetic energy. The exit and entrance slits were separated by an angular distance of $\pi\sqrt{2}$ radians which is a condition for optimum resolution. Construction of the instrument used in this work was done under the supervision of Bemis. Details are presented elsewhere.[2]

The spectrometer measures the intensity of electrons emitted as a function of their kinetic energy. Peaks occur at energies characteristic of elements present in the irradiated sample. The kinetic energy of a characteristic peak of an element is given by:

$$E_k = E_{xr} - E_B$$

where E_k is the kinetic energy measured for the peak, E_{xr} is the energy of the x-ray photons impinging on the sample, and E_B is the binding energy that

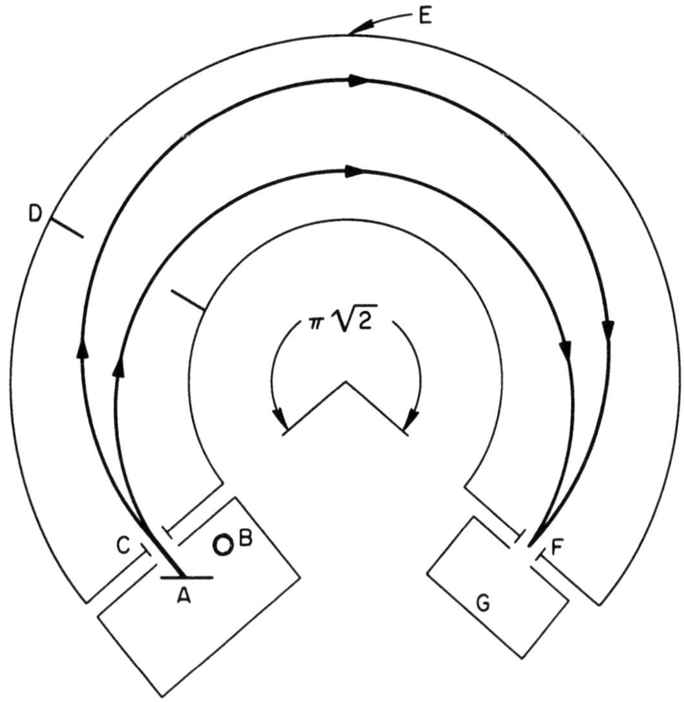

Double Focusing Iron-Free Magnetic Electron Spectrometer

Fig. 1

held the electrons within the atoms from which they were ejected. Aluminum K_α radiation was used in these studies and is commonly used by other workers. The energy of the photons was 1487 eV; binding energies ranged from 100 to 500 eV; therefore, kinetic energies were typically 900 to 1400 eV.

The ejection of electrons from specimens causes them to be charged positively if they are insulators. This reduces the measured kinetic energies of peaks. A convenient way of correcting for this is to use the carbon peak as a reference. Adsorbed carbon compounds are almost always present on specimens. For a given specimen, the charging shifts the carbon peak the same amount as the peaks for the other elements.

Peaks in photoelectron spectra have widths of 1.5 to 3.0 eV at one-half their maximum values. This broadening is due to the width of the excitation source and to variations in potentials on the surface of the specimen rather than to resolution limitations of the spectrometer. The reason for the potential variations is not clear. The kinetic energies of peaks can be measured to one-tenth their widths (0.2 eV).

The heights of peaks are proportional to the concentrations in the specimen of their corresponding elements. Thus the method allows quantitative as well as qualitative analysis. It may be used in analysis for all elements of atomic number higher than 2.

Chemical Shifts

The charge induced on an atom by other atoms in chemical combination with it causes changes in its inner shell binding energies. In general, atoms of higher oxidation states have higher binding energies. The spectrum of an element present in a specimen in more than one oxidation state will have more than one peak if the binding energy differences for the oxidation states are large. If the binding energy differences are small, the spectrum will appear only as a broadened peak. Chemical shifts for most of the light elements have been reported.[1] Shifts in heavier elements have been observed also. All elements have not as yet been studied, but there is reason to believe that they all should exhibit measurable chemical shifts. Chemical shift plots, that is, graphs of binding energy differences of elements in different chemical states vs. their effective electron densities resulting from their environments, have been constructed. Charge densities on atoms can be calculated by quantum mechanical methods or by electronegativity considerations. The charge densities surrounding atoms of an element (more crudely, their oxidation states) in an unknown compound can be evaluated by comparing the binding energies measured for the element with those on chemical shift plots.

In the work presented here, the chemical shift effect will be used to determine the oxidation states of sulfur in smoke and fly ash specimens. This technique has also been applied to another pollution problem involving arsenic[3] and to problems of biological interest.[4]

EXPERIMENTAL RESULTS

Electron kinetic energies of the spectra in this work were all referenced to those of the carbon peak to correct for specimen charging effects. Oxidation states of sulfur in the specimens of this work were determined by comparing chemical shifts in binding energies with those measured for known compounds. The binding energy of sulfur for many oxidation states has been tabulated, and plots of chemical shifts vs. oxidation number and calculated charge have been made.[1]

Specimens of smoke particles from coal burned in a home fireplace were collected on filter paper and examined by photoelectron spectroscopy. Electrons emitted from 2 p energy levels in sulfur were studied. The spectrum is shown by the lower curve in Fig. 2. The multiple peaks demonstrate the chemical shift effect. The more narrow peak, having the highest kinetic energy (lowest binding energy), indicates a single species of sulfur in a reduced state, probably hydrogen sulfide or a mercaptan. Deconvolution of the broader peak at higher kinetic energy shows that there are two species of higher oxidation state--sulfite and sulfate. The reference spectrum, uppermost in Fig. 2, is for electrons emitted from 2 p levels of sulfur in pure potassium sulfate.

The sulfur spectrum (2 p level) of the fly ash particles is also a composite of peaks from more than one chemical state. Both peaks are probably of sulfate or adsorbed SO_3 species, since their binding energies are much higher than those for sulfite states. Perhaps variations in the nature of the substrate

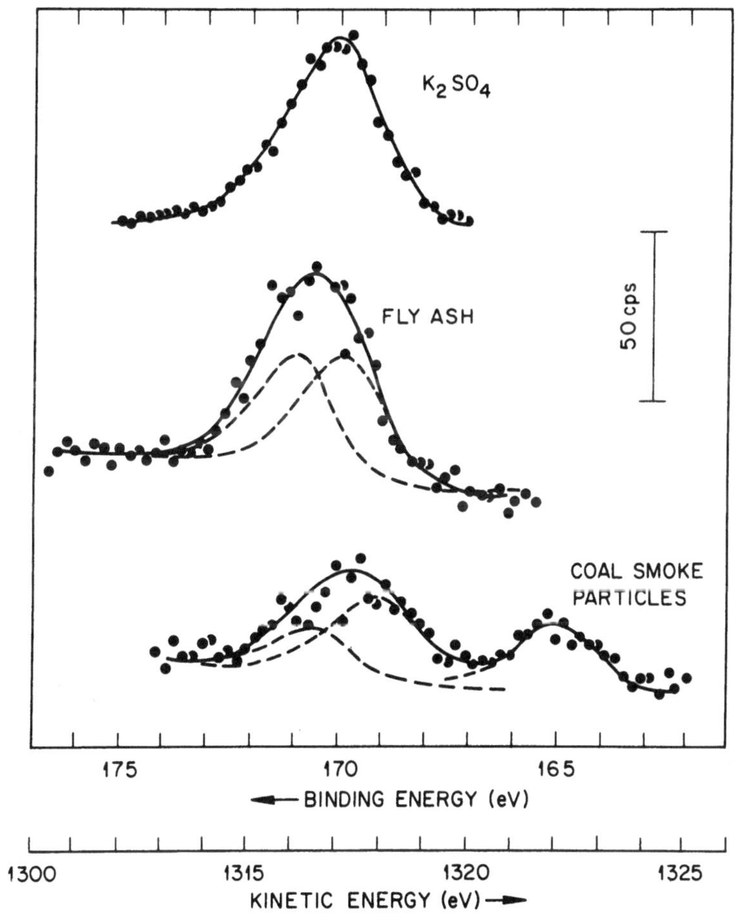

Photoelectron Spectra of Sulfur on Various Substrates

Fig. 2

on which the sulfur is adsorbed cause differences in its binding energy; further study is necessary before more specific conclusions can be made.

Figure 3 shows the sulfur spectra (2 p energy levels) of metal oxides that have been exposed to sulfur dioxide. Exposure to the SO_2 was done at atmospheric pressure and room temperature with air contact allowed. For the two transition metal oxides, Fe_2O_3 and MnO_2, the spectra indicate that sulfate is the only species present. Apparently the oxidation of SO_2 to sulfate occurred very rapidly on the transition metal oxide surfaces. The spectra for sulfur

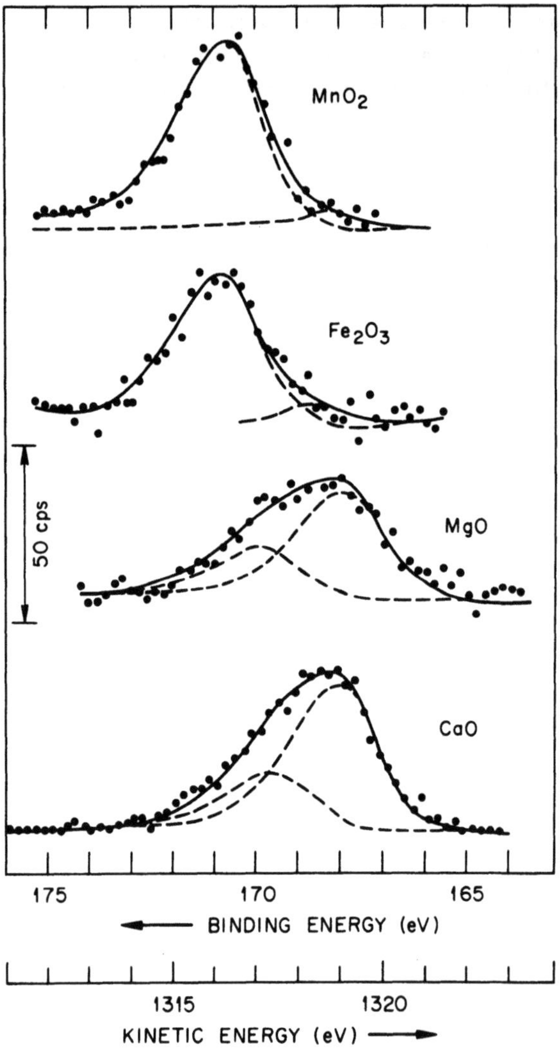

Photoelectron Spectra of Sulfur on Various Substrates

Fig. 3

on the alkaline earth oxide surfaces indicate a mixture of sulfate and sulfite states. The peaks are too broad to be explained by a single sulfur species. Deconvolution of the peaks shows that the sulfite species predominates. Oxidation to the sulfate occurred very slowly on the alkaline earth oxides.

The fly ash and smoke particle specimens were analyzed for total sulfur content by combustion analysis (Leco commercial analyzer). The fly ash specimen was the same one used for measuring the spectrum in Fig. 2. The smoke particle specimen was not the same as that for Fig. 2, but it was prepared under similar combustion conditions. Total concentrations of sulfur were 0.15% in the fly ash and 0.95% in the smoke particles.

DISCUSSION AND CONCLUSIONS

The results of this study have shown that photoelectron spectroscopy can be used to detect sulfur and determine its chemical state on the surfaces of field specimens of particulate matter. Thus we have potentially a new means of surveillance that will allow for more thorough assessments of pollution hazards. The present monitoring procedures, which measure gas and solid components separately, may not be sufficient. For example, situations might arise where gas analysis alone would indicate safe levels of SO_2, while in reality there might be unacceptably high amounts of sulfur compounds being transported on particle surfaces.

The sulfur peak for the fly ash specimen of Fig. 2 is remarkably strong, considering that the total sulfur concentration, as measured by combustion analysis, was only 0.15%. This suggests that the sulfur was present mostly as a layer adsorbed on the surface of the particles rather than as an homogeneous component. The photoelectron spectroscopy technique is very sensitive for surface analysis; as little as one monomolecular layer can be detected.[1] The intensity of the fly ash sulfur peak shows that specimens of much lower concentration (10-100 ppm) can be studied. A proper comparison of the strength of the sulfur peaks for the smoke particles with the combustion analysis results cannot be made since different specimens were analyzed by the two techniques. It is probable, however, that the photoelectron spectroscopy technique will be very sensitive for analyzing specimens of this type also.

An advantage of this technique is its specificity in element identification. The binding energies of the inner electron shells of sulfur are sufficiently unique that it is extremely unlikely that other elements will interfere. There is no question that the peaks in the coal smoke particles and fly ash spectra (Fig. 2) are of sulfur. It is clear that this technique should be field tested further and, if possible, developed into a standard method of analysis.

Probably this method can also be used to analyze for aggressive compounds of other elements, such as nitrogen.

In smoke pollution research it is desirable to know what variables affect the formation of aggressive chemical species and their interaction with the solid components. The results in Fig. 2 suggest that variables in the furnace, such as combustion efficiency, are important. A high proportion of the sulfur compounds on the surfaces of the coal smoke particles produced in the home furnace were in the more reduced sulfide and sulfite states. This was probably caused by the low combustion efficiency conditions which produced a reducing atmosphere of CO. The sulfur species on the fly ash specimen were all in the more highly oxidized sulfate form. This specimen was taken from a local power plant where combustion efficiency is carefully controlled and the atmospheres in the firebox and precipitators are more highly oxidizing than in home furnaces. Speculations about why the fly ash was in the sulfate form must be made with reservations, however. The specimens analyzed in this study were several months old, and it is possible that sulfite or other reduced forms were initially present but were oxidized by air exposure.

There are probably many variables in the downwind smoke plume, such as humidity, temperature, light exposure, and composition of the particulate matter, that also affect the chemical species adsorbed on the surfaces of the solid components. The results in Fig. 3 show that photoelectron spectroscopy can be used to investigate some of these variables. It is shown that the particulate matter composition is very important. Transition metal oxides caused rapid oxidation of adsorbed sulfur dioxide to sulfate, whereas on alkaline earth oxides the oxidation was much slower.

Information such as that in Fig. 3 is also useful in sulfur dioxide abatement research. Several methods are under consideration which involve adsorption on solid surfaces, and there are many variables that affect their efficiency. Photoelectron spectroscopy is a technique by which the effects of many of these variables can be quickly determined.

ACKNOWLEDGEMENT

The authors are very grateful to C. E. Bemis and O. L. Keller, Transuranium Laboratory, ORNL, for providing the iron-free magnetic spectrometer that was used in this work. The help of W. R. Laing, who supervised the combustion analysis work, is also appreciated.

REFERENCES

1. K. Siegbahn et al., ESCA: Atomic, Molecular and Solid State Structure Studied by Means of Electron Spectroscopy (Almquist, Uppsala, Sweden, 1967).

2. C. E. Bemis, "The $\pi\sqrt{2}$ Double Focusing Electron Spectrometer," Chemistry Division Annual Progress Rept., May 20, 1968, ORNL-4306.

3. L. D. Hulett and T. A. Carlson, "Measurements of Chemical Shifts in the Photoelectron Spectra of Arsenic and Bromine Compounds," Applied Spectroscopy 25, 33 (1971).

4. L. D. Hulett and T. A. Carlson, "Analysis of Compounds of Biological Interest by Electron Spectroscopy," Clinical Chemistry 16, 677 (1970).

COMPARISON OF METHODS FOR THE DETERMINATION OF NITRATE. DETERMINATION OF NITRATE THROUGH REDUCTION

C. R. Sawicki * and F. Scaringelli**

Wing Q, Room 301, Research Triangle Park,
North Carolina* and 3820 Merton Drive,
Raleigh, North Carolina**

ABSTRACT

A variety of direct and reduction methods for the determination of nitrate are compared in terms of simplicity, sensitivity, precision and selectivity. Stress was placed on the development of reductive procedures wherein nitrate and/or nitrite could be determined by a slight modification in the basic procedure. Of the large group of reducing agents the hydrazine-cupric ion system was found to be especially attractive. Nitrite formed in the reaction was capable of being determined by a large variety of reagents. The parameters in the reductive and chromogen-formation procedures are discussed at length.

PHOTOMETRIC DETERMINATION OF POLYPHENOLS IN PARTICULATE MATTER

E. Sawicki and M. Guyer

Wing Q, Room 301, Research Triangle Park

North Carolina 27709

ABSTRACT

The evidence indicates that polyphenols could play a role in our understanding of the physiological activities of the aerocarcinogens and aeroallergens. Quantitative extractions of polar organic molecules containing hydroxyl groups from complex mixtures containing inorganic and organic components was found to be an extremely difficult problem that will necessitate many studies before the best solutions will be found. In this preliminary study it was necessary to investigate the extractability of polyphenols from various types of particles. The procedure of analysis then had to be adapted to the solvent chosen for extraction. On this basis a group of colorimetric and fluorimetric methods were developed for the analysis of the polyphenols. The methods were applied to the analysis of airborne particulates, airborne pollen and fungi and dusts of various kinds.

AN EVALUATION OF ATOMIC ABSORPTION AND FLAME EMISSION SPECTROMETRY FOR AIR POLLUTION ANALYSIS

T. C. Rains, T. A. Rush, and O. Menis

Analytical Chemistry Division

National Bureau of Standards, Washington, D. C. 20234

ABSTRACT

In many cases the concentration of metals in particulate matter is very low and therefore requires the ultimate in sensitivity. Frequently, analytical methods used lack the desired precision and accuracy. Also, absence of specificity inherent in many techniques can lead to complications. Atomic absorption and flame emission are being evaluated for the determination of selected elements encountered in atmospheric particulate. As a laboratory control sample a large volume of atmospheric particulate has been obtained and this material is being evaluated by various analytical competences. The discussion will include a description of the instrumentation, dissolution of sample and analytical results. Data are also presented on the effects of interferences and some techniques for their elimination or control.

ANALYSIS OF THE AEROCARCINOGEN CONGLOMERATE

E. Sawicki

Wing Q, Room 301, Research Triangle Park

North Carolina 27709

ABSTRACT

It is now becoming evident that in the analysis for any physiologically active component of a mixture found in the human environment analytical data is also needed on cofactors in this and other mixtures present in the environment, these cofactors having a definite effect on the activity of the physicogen of interest. In this paper we will discuss the current status of the methodology for the analysis of the aerocarcinogen conglomerate. A critique of the methodology for the carcinogens, anticarcinogens, cocarcinogens, precarcinogens, irritants and enhancers will be presented. The future needs in this field will be discussed.

MICROWAVE SPECTROMETRY AS AN AIR POLLUTANT

ANALYSIS METHOD

Edgar A. Rinehart

Physics Department, University of Wyoming

Laramie, Wyoming 82070

ABSTRACT

The high resolution and resulting high specificity of molecular rotational resonance (MRR) spectrometry, along with the high sensitivity for certain polar compounds suggests its use for the detection and measurement of trace quantities of these compounds in samples of polluted air.

Methods of intensity measurement under conditions of transition saturation result in a signal which is linear with partial pressure of the absorbing species over a range of from 100% to a few parts per billion for some compounds.

These methods will be discussed, and results of quantitative measurements of these compounds in air samples will be presented.

Some new developments in instrumentation making the method more convenient and less expensive will also be discussed.

BIOLOGICAL DEGRADATION OF TOXIC POLLUTANTS

G. G. Guilbault, S. S. Kuan, M. H. Sada,
W. Hussein, and S. Hsiung

Department of Chemistry, Louisiana State University
New Orleans, Louisiana 70122

ABSTRACT

Biological degradation of toxic pollutants promise success for the purification of the environment. Enzymes are specific catalysts which are able to degrade specifically certain chemical compounds. Organophosphorus pesticides, for example, can be degraded by the enzymes lipase and bacterial and animal cholinesterases to non-toxic fragments. The enzymes rhodanase and cyanide injectase are capable of converting the toxic HCN to non-toxic product. Carbamase catalyzes the specific hydrolysis of carbamates. Hydrocarbon oxidases isolated from soil bacteria are capable of oxidizing selective hydrocarbons to CO_2 and water. The use of these and other enzymes systems in the identification of toxic substances in air and in the purification of the environment will be discussed. Methods of inducing specific enzymes for the degradation of selective pollutants will be discussed as will various immobilization techniques for holding the enzyme to retain its activity over long periods of time. Spectroscopic and electrochemical methods for monitoring the identification and degradation reactions will be discussed, and the possible application of such techniques to large-scale anti-pollution devices will be covered.

INDEX

Acrolein, 148
Aeroallergens, 192
Aerocarcinogens, 191, 192
Aerometric data, 25
Air Quality Control Region, 20
Aldehydes, 2
Ammonia, 2, 4, 164
Amperometry, determination of ozone (oxidant), 109-112
Analysis range, 89
Anticarcinogens, 191
Antimony, 148
Arsenic, 4, 148, 182
Atomic absorption spectroscopy, 132, 194

Benzo(a)pyrene, 4
Beryllium, determination of, 138, 148
Binding energy, 180, 181
Biological degradation, 190

Cadmium, determination of, 139
Calibration, 36
  of ozone-oxidant instruments, 104
  of SO_2 instruments, 87, 91
Carbon monoxide, 4, 5, 22, 67, 76, 148
Carcinogens, 191
Charge density, 182
Chemical shifts, 182
Chemiluminescence, 15
  ozone monitor, 85, 91, 92
CHESS, 41
  -CHAMP, 45
  health indicators, 44
  monitoring, 46
Chloride, 48
Chlorine, 148
Chromatography, 15, 60, 83, 94, 165, 166
Chromium, determination of, 139
Cobalt, determination of, 139-140
Cocarcinogens, 191
Colorimetry
  determination of ozone (oxidant), 109-111, 113-115
  determination of SO_2, 83, 94
Combustion efficiency, 186
Community Health and Environmental Surveillance Studies, 41
Conductometry, determination of SO_2, 83, 94
Continuous Air Monitoring Program (CAMP), 5
Copper, determination of, 140
Coulometric detector for SO_2, 83, 94
Criteria, 19
  document, 7, 21

Diurnal patterns, 126
Dustfall, 148

Electron spectrometer, 181
Emission sampling, 153
Enhancers, 191

Enzyme degradation, 190
ESCA, 180

Fall time, 89
Filter, activated charcoal and soda lime, 110
Flame photometric detector, 15, 83, 94, 164, 166, 194
Fluoride, 2, 4, 9, 148
Fly ash, 179
Formaldehyde, 154

Gas chromatography, 165, 166
Glass filters, 133

Halogens, 164
High-volume sampler, 133
Human pollutant burden, 49
  pattern, 50
Hydrocarbons, 2, 5, 22, 67, 76, 78, 148
Hydrogen flame ionization detector, 164, 165
Hydrogen sulfide, 2, 84, 102, 148, 164, 183

In situ analysis, 180
Index
  Kovats, 165
  pesticide exposure, 57
  Pindex, 65
Intersociety Committee on Manual of Methods for Ambient Air Sampling and Analysis, 143
Iodine-131, 148
Iodometry, determination of oxidant, 109, 110
Iron, 132, 148
Irritant, 191

Kovats index, 165

Lag time, 89
Lead-210, 148
Lead, determination of, 140

Manganese, 132, 148
Mercaptan, 104, 148, 183
Methylene blue determination of H_2S, 102
Microwave spectroscopy, 189
Minimum detectable concentration, 89
Molecular rotational resonance (MRR) spectrometry, 189
Molybdenum, 148
Monitor, human tissues as, 49

National Aerometric Data Bank (NADB), 7, 14
National Aerometric Data Information Service (NADIS), 11-13
National Air Sampling Network (NASN), 3, 6
Nederbragt detector, 91, 104
Network
  Hi-Vol total suspended particulates, 8
  total mercury, 9
  membrane filter and precipitation, 9
  aerometric data, 25
Nickel, determination of, 140
Nitrate, 148, 193
Nitric oxide, 5, 110
Nitrite, 193
Nitrogen dioxide, 5, 109, 110, 112, 113, 148
  colorimetric analyzer, 115
  interference equivalent, 114, 129
Nitrogen oxides, 2, 15, 22, 67, 76, 78, 148

Odor, 155
  acceptability (hedonic tone), 156, 167
  detectability, 156, 167, 170, 173
  perception (sensation), 155, 156, 163

Odor (continued)
quality, 156, 167, 175
stimulus, 158, 163
supraliminal intensity, 156, 167, 170, 172
thresholds, 158, 168, 173
Odorous emissions, 155
Olfactory fatigue, 171
Olfactory sensor, 163
Olfactometer, 158
Organophosphorus pesticide, 19
Oxidant, 2, 5, 22, 76, 78, 109, 148
analyzers, 91
Oxidant state, 179, 180, 182
Oxide, alkaline earth, 185, 186
Oxide, transition metal, 184, 186
Ozone, 84, 164
analyzer, 91
determination of 109, 110
generation, 116

Panels, 169
Particulate matter, 2, 22, 67, 73, 148, 179, 186, 192
constituents of, 10
Permeation tube, 88
Peroxyacetyl nitrate (PAN), 100, 148
Pesticides, 50, 52
exposure index, 57
Phenolic compound, particulate, vapor phase, 148
Photoelectron spectroscopy, 179, 180
PINDEX, 65
Plutonium, 148
Polynuclear hydrocarbons, 148
Polyphenols, 192
Precarcinogen, 191
Psychophysical method, 156, 167

Radioactivity, alpha and beta, 148
Regener ozone analyzer, 90
Ring oven technique, 132, 135-141
Rise time, 89

Sampling, trace metals, 138
SAROAD system, 6, 11, 152
Selenium, 131, 148
Sensory detectability, 165
Sensory methods, 155, 156, 163
Smoke particles, 179, 182
Standards, 53, 68
emission, 23
performance, 23
primary, 21, 71, 72
secondary, 22, 71
Statistical tests, 121, 124, 125, 128
Strontium-89 and 90, 148
Sulfate, 179, 183
Sulfide, 164, 167
Sulfite, 183
Sulfur compounds, 83
Sulfur dioxide, 5, 22, 83, 84, 109, 110, 129, 148, 164, 179, 180
interference in oxidants analysis, 106
sensors, 95
scrubber for removal of, 110
Sulfur oxides, 67, 74
Sulfur trioxide, 183
Surface analysis, 185
Surveillance, 1, 6, 41
Suspended particulate matter, 148

Tape sampler, sequential, 139
Thiol, 164
Tolerance factor, 68
Trace metals, 131-141
Trigeminal sensor, 163, 164
Tritium, 148

West-Gaeke method, 83, 85, 94

X-ray photon, 180
source, 180

Zinc, determination of, 141

MIX
Papier aus verantwortungsvollen Quellen
Paper from responsible sources
FSC® C105338

If you have any concerns about our products,
you can contact us on
ProductSafety@springernature.com

In case Publisher is established outside the EU,
the EU authorized representative is:
Springer Nature Customer Service Center GmbH
Europaplatz 3, 69115 Heidelberg, Germany

Printed by Libri Plureos GmbH
in Hamburg, Germany